Studies in Logic
Volume 114

Logic and Language
New Frontiers in Analysis and Interpretation

Volume 104
Argument, Sex and Logic
Dov Gabbay, Gadi Rozenberg and Lydia Rivlin

Volume 105
Logic as a Tool. A Guide to Formal Logical Reasoning
Valentin Goranko

Volume 106
New Directions in Term Logic
George Englebretsen, ed

Volume 107
Non-commutative Algebras. Pseudo-BCK Algebreas versus m-pseudo-BCK Algebras
Afrodita Iorgulescu

Volume 108
Semitopology: decentralised collaborative action via topology, algebra, and logic
Murdoch J. Gabbay

Volume 109
The Cognitive Dimension of Social Argumentation. Proceedings of the 4th European Conference on Argumentation, Volume I. Fabio Paglieri, Alessandro Ansani and Marco Marini, eds.

Volume 110
The Cognitive Dimension of Social Argumentation. Proceedings of the 4th European Conference on Argumentation, Volume II. Fabio Paglieri, Alessandro Ansani and Marco Marini, eds.

Volume 111
The Cognitive Dimension of Social Argumentation. Proceedings of the 4th European Conference on Argumentation, Volume III. Fabio Paglieri, Alessandro Ansani and Marco Marini, eds.

Volume 112
Implicative-groups vs. Groups and Generalizations. Second Edition
Afrodita Iorgulescu

Volume 113
Meaning as a Set-theoretic Object. A Gentle Introduction to the Ideas Behind Formal Semantics
Jaroslav Peregrin

Volume 114
Logic and Language: New Frontiers in Analysis and
José David García-Cruz, ed

Studies in Logic Series Editor
Dov Gabbay   dov.gabbay@kcl.ac.uk

Logic and Language
New Frontiers in Analysis and Interpretation

Edited by
José David García-Cruz

© Individual authors and College Publications, 2025
All rights reserved.

ISBN 978-1-84890-493-4

College Publications
Scientific Director: Dov Gabbay
Managing Director: Jane Spurr

http://www.collegepublications.co.uk

Cover prepared by Laraine Welch

All rights reserved. No part of this publication may be reproduced, stored in a retrieval system or transmitted in any form, or by any means, electronic, mechanical, photocopying, recording or otherwise without prior permission, in writing, from the publisher.

Contents

Introduction 1
José David García-Cruz

Logical Analysis of Language and Physicalism in Rudolf Carnap 9
Fernando Huesca Ramón

Argumentation, Images, and the Concept of Information . . . 19
Francisco Montes and Fernando Huesca Ramón

Topology and Peirce's Existential Graphs: The Case of the
  Möbius Strip 35
Paniel Reyes Cárdenas and José Martin Castro

An Ambivalent Interpretation of the *dictum de omni* in *Prior
  Analytics* B, 22, 68a16-68a25 47
Luca Gili

Hylomorphism and the Platonization of Logic 59
Karel Sebela

The Ontological Argument and the Modal Square of Opposition.
  A Dialogue between a Theist and an Atheist. Is there a
  Possible Reconciliation? 73
Juan Campos

A Logic for Prophetic Conditionals: Perfect Prophecy and
Prophetic Intuition 87
José David García-Cruz

Towards Non-Deductive Term Functor Logic 113
José Martín Castro

Many-Valued Oppositions 123
Yessica Espinoza Ramos

Logic-Sensitivity and Bitstring Semantics: From Squares to
Hexagons of Opposition 147
Lorenz Demey and Stef Frijters

Introduction

José David García-Cruz

1 The Context of the Book

This volume, *Logic and Language: New Frontiers in Analysis and Interpretation*, emerges from a series of academic endeavors focused on the intricate relationship between logic and language. Between 2021 and 2023, a group of online events—most notably the *Syllogism and Argumentation* workshops[1]—brought together scholars from diverse backgrounds to delve into topics ranging from syllogistic reasoning to the broader implications of argumentation in philosophical discourse.

The inaugural workshop, held in June 2022, facilitated exploratory discussions on the distinctions and intersections between syllogism and argumentation. Participants examined foundational questions, such as the unique characteristics that differentiate syllogistic structures from general argumentative forms, and whether argumentation could be viewed as an extension of syllogistic reasoning. Sessions featured presentations on topics like existential import in syllogistics, sortal logic, and historical analyses of Aristotle's methodologies.

Building upon the success of the first workshop, *Syllogism and Argumentation II* convened in July 2023, further expanding the dialogue to encompass the philosophical, logical, and mathematical foundations of syllogistics and argumentation. This second event continued the tradition of fostering interdisciplinary discussion, highlighting the evolving nature of logical analysis in contemporary scholarship.

Both workshops were organized as part of two research projects funded by the John Templeton Foundation, under the direction of the editor of this volume: *"The Dynamic Logic of Religious Conversion"* and *"The Logic of Prophetical Conditionals: Prophetical Language, Divine Communication and Human Freedom"*. These initiatives provided the institutional and intellectual framework for the collaborative work presented in this book.

---

[1] Fo more details: Syllogism and Argumentation and Syllogism and Argumentation II

2 Motivations and Rationale

The present volume brings together a series of contributions that explore the intricate and multifaceted relationship between language and logic. Rather than offering a historical or disciplinary overview, the chapters collected here aim to address a conceptual question: in how many ways can the interplay between language and logic be understood, formalized, or problematized?

This guiding question is approached through a variety of perspectives—from ancient logic to modal logic in theology, from logical geometry to many-valued logics, and from term-functor logic to logical analysis of language. What unites these otherwise diverse approaches is a shared conviction that the bond between language and logic is not merely instrumental or superficial, but constitutive of certain forms of rationality. The aim of this volume, therefore, is to trace a conceptual cartography of this bond, inviting the reader to reflect on the ways in which logic both shapes and is shaped by linguistic structures and philosophical inquiry.

What sets this volume apart from existing literature is not only the breadth of topics it addresses, but the originality of its philosophical method and style. Among the most distinctive contributions is a Socratic-style dialogue in which the interlocutors are not merely literary constructs, but scholars specialized in modal logic. This creative reappropriation of the Platonic tradition invites readers to engage with contemporary logical issues through a form that privileges inquiry, critical exchange, and conceptual precision. Far from being an exercise in nostalgia, this dialogical format reveals new ways of thinking philosophically about logic, rekindling a genre that is as intellectually demanding as it is pedagogically effective.

In addition to its methodological diversity, the book offers original contributions to active areas of research, particularly in the study of logical oppositions and non-classical logics. Two chapters stand out in this regard: one explores how recent developments in logical geometry can be extended to non-classical systems, while the other addresses the problem of logic sensitivity within modal frameworks. These chapters not only advance specific debates within their respective fields, but also exemplify the volume's broader ambition—to provide conceptual tools that are both analytically rigorous and open to interdisciplinary dialogue.

Because of its range in both content and style—from accessible philosophical dialogues to technically sophisticated essays—this volume serves as a rich resource for multiple audiences. It is suited for students and newcomers seeking orientation in the interplay between language and logic, as well as for established scholars in logic, the history of logic, and the philosophy of language. Whether one approaches the text with pedagogical, historical, or theoretical interests, the volume invites sustained engagement with fundamental questions concerning the nature, structure, and function of logical thought as it emerges through language.

Taken together, the contributions in this volume do not aim to impose a unified theory or framework, but rather to chart a constellation of approaches that illuminate the dynamic intersections between language and logic. The result is a polyphonic account, one that reflects both the diversity of logical traditions and the shared commitment to exploring the conceptual terrain where linguistic meaning and logical form converge. In this sense, the volume resists disciplinary closure, inviting instead a dialogue that crosses boundaries between analytic rigor and philosophical imagination.

Ultimately, the motivation behind this collection is neither merely historiographical nor purely technical. It is animated by the conviction that logic is not an abstract enterprise removed from linguistic life, but a mode of thinking that both shapes and is shaped by the languages we use to reason, argue, and inquire. By gathering voices from different traditions and levels of specialization, this book aims to foster not only understanding, but also curiosity—an openness to the evolving landscape of logical thought as it continues to engage with the deepest structures of language and meaning.

3 Structure of the Chapters and Organization

The volume is divided into ten chapters, organized thematically to reflect the diversity and progression of topics addressed. The first chapters focus primarily on issues related to the philosophy of language, followed by contributions that engage with problems from ancient logic. The middle section of the book explores intersections between logic and the philosophy of religion, while a subsequent chapter delves into term logic. The volume concludes with two chapters dedicated to topics in logical geometry, providing a coherent structure that guides the reader through classical and contemporary perspectives in formal reasoning.

The opening chapter, authored by Fernando Huesca Ramón, revisits the intellectual legacy of Rudolf Carnap, one of the central figures of the Vienna Circle and a staunch advocate of logical positivism. Through a careful historical and conceptual analysis, the chapter explores Carnap's enduring commitment to the logical analysis of language and its implications for a physicalist conception of the world. It traces key moments in Carnap's philosophical trajectory—from his critical engagement with metaphysics and his polemics against Heidegger, to the debates surrounding the analytic/synthetic distinction and the epistemic status of scientific language. The author highlights the nuanced position Carnap occupies between constructivist and naturalist approaches, ultimately arguing for a more balanced interpretation of his philosophical project—one that resists simplistic dichotomies, such as the oft-invoked opposition between Carnapian logical syntax and Kuhnian historicism.

In the second chapter, Francisco Montes examines the role of images in non-verbal communication and argumentation, drawing attention to their ambiguous status as conveyors of information. Anchored in a vivid anecdote involving gestural communication in a conference setting, the chapter illustrates how meaning can be inferred from actions and visual cues even in the absence of spoken language. Montes explores the philosophical implications of such cases, focusing in particular on the capacity of images—such as diagrams, symbols, or icons—to serve argumentative functions. Building on Barceló-Aspeitia's thesis that images can directly express the propositional content of premises and conclusions, the author engages in a critical discussion of when and how images succeed—or fail—in transmitting information. By appealing to Dretske's semantic theory of information, the chapter offers a refined account of the conditions under which images contribute meaningfully to argumentative discourse, and the limits of their epistemic reliability.

Chapter three, authored by Paniel Reyes Cárdenas and José Martín Castro Manzano, offers an original and provocative contribution at the crossroads of logic, topology, and Peircean semiotics. Centering on the Möbius strip—a paradigmatic example of a non-orientable surface—the chapter investigates its potential as a diagrammatic space within Charles Sanders Peirce's system of existential graphs. The authors argue that such topological structures can accommodate the coexistence of a proposition and its negation without entailing contradiction, thereby providing a novel approach to classical paradoxes such as the "round square." Two interpretative strategies are proposed: one grounded in set-theoretical considerations, and another that leverages the expressive resources of existential graphs to reframe the paradox. The chapter also revisits Arnold Oostra's underexplored contributions to the field, underscoring the diagrammatic and cognitive virtues of Peirce's graphical syntax. In combining topological insights with logical analysis, the authors chart a promising path for visual reasoning and the representation of logical anomalies.

In chapter four, Luca Gili revisits the long-standing debate surrounding the interpretation of the Aristotelian *dictum de omni et de nullo*, particularly as it appears in *Prior Analytics B.22 (68a16–68a25)*. This *dictum*, traditionally regarded as a cornerstone of syllogistic reasoning, has been the subject of divergent scholarly readings, most notably those labeled "orthodox" and "heterodox" by Jonathan Barnes and Michael Frede, respectively. Gili challenges the binary framing of this interpretive dispute by advancing the thesis that Aristotle's own position may in fact be ambivalent, accommodating both readings depending on the logical and textual context. Through a close philological and philosophical analysis, the chapter offers a nuanced reassessment of a key Aristotelian principle, shedding light on its flexible function within the *Analytics* and its implications for understanding the foundations of ancient logic.

Chapter five, authored by Karel Šebela, offers a provocative historical and philosophical investigation into the role of hylomorphism in the development of logic. While logic has traditionally privileged form over content—culminating in the formalist paradigm that dominates contemporary discourse—Šebela challenges the assumption that this emphasis on form originates with Aristotle. He contends instead that logical hylomorphism, the distinction between matter and form in logic, represents a crucial turning point in the "Platonization" of logic. By tracing the emergence and evolution of this distinction, particularly through Kant's formulation of logic as a purely formal discipline, Šebela argues that hylomorphism entails a specific theory of concepts, without which its logical function cannot be sustained. This chapter thus situates the form–matter dichotomy at the intersection of historical developments and deep conceptual commitments, revealing the metaphysical underpinnings of modern formal logic.

In chapter six, Juan Manuel Campos Benítez stages a philosophically engaging dialogue between a theist and an atheist who, despite their opposing metaphysical commitments, share a common logical framework. Centered on the ontological argument for the existence of God, the discussion makes use of the modal square—and more broadly, the octagon—of opposition as a formal tool to map and assess the logical structure of competing claims. By embedding classical theological debate within a rigorous modal framework, Campos Benítez invites the reader to consider whether reconciliation is possible not through persuasion, but through the shared architecture of modal reasoning. This chapter thus illuminates the potential of formal logic to mediate deep existential disputes.

Chapter seven, authored by José David García Cruz, presents BTEP, a novel multimodal logic developed to formalize reasoning about prophetic conditionals. Drawing on Ockham's account of divine foreknowledge and Aquinas' distinction between prophetic intuition and perfect prophecy, the chapter introduces a three-dimensional semantic framework grounded in a branching-time model.

The system is designed to address the epistemic complexity of prophetic agents, whose discourse blends modal necessity with contingent truth conditions. Central to the inquiry is the role of the prophet as an epistemic agent mediating divine and human perspectives on future contingents. Standard modal systems fall short in capturing such epistemic dynamics, prompting the need for BTEP's enriched structure. The chapter concludes with two case studies—on the prophecy of Nineveh and on Aquinas' taxonomy of prophecy—illustrating the logic's capacity to represent both theological and epistemological dimensions of prophetic reasoning.

Chapter eight, authored by José Martín Castro Manzano, advances the development of Term-Functor Logic (**TFL**) by proposing an extension capable of accommodating non-deductive inference. While **TFL** adheres to the syntactic architecture of traditional Aristotelian logic, its integration of inductive and abductive reasoning remains an open challenge. This chapter addresses that gap by incorporating elements from Non-Axiomatic Logic—specifically designed to handle non-deductive reasoning—into the **TFL** framework via a tableaux method. The proposed hybrid system yields a preliminary tableau-based approach to modeling non-deductive inferences within a term-logical context. This contribution not only strengthens the relevance of **TFL** to the Aristotelian tradition but also paves the way for a broader inferential capacity in formal systems that retain term-centric syntax.

Chapter nine, by Yessica Espinoza Ramos, examines the viability of preserving the classical Aristotelian square of opposition within a range of non-classical logical systems. Focusing on five many-valued and paraconsistent logics—**K3**, **Ł3**, **LP**, **RM3**, and **FDE**—the chapter assesses whether the traditional oppositional relations (contradiction, contrariety, subcontrariety, and subalternation) remain intact under conditions of indeterminacy or inconsistency. The analysis reveals that only **LP** and **RM3** fully sustain the classical structure, while the others fail to do so either partially or completely. In response, the chapter introduces a generalized semantic framework based on positive and negative satisfaction, allowing for a redefinition of oppositional relations compatible with these alternative logics. This generalization culminates in an enriched aristotelian relations that both preserves and extends the classical model, offering a unified structure across multiple logical systems.

Chapter ten, authored by Lorenz Demey and Stef Frijters, advances the study of logical geometry by exploring the complex interplay between logic-sensitivity and bitstring semantics within Aristotelian diagrams. Building upon previous work that examined this interaction in the square of opposition, the chapter extends the investigation to hexagons of opposition. The authors argue that while the interaction remains tractable and elegant at the purely combinatorial level—where bitstrings are treated as abstract sequences—it becomes significantly more intricate when those bitstrings are interpreted semantically, as formulae within specific logical systems. The chapter thus highlights both the promise and the limitations of bitstring semantics as a tool for representing logic-sensitive structures, advocating for a broader and more nuanced approach to diagrammatic opposition.

The contributions gathered here reveal a fruitful diversity of perspectives on logic and philosophy of language, ranging from classical and modal frameworks to non-deductive and non-classical extensions. The thematic division adopted here not only reflects the internal coherence of each cluster of topics, but also highlights the conceptual bridges that connect them. The richness of these discussions confirms the enduring relevance and versatility of logic in contemporary philosophical and formal investigations.

LOGICAL ANALYSIS OF LANGUAGE AND PHYSICALISM IN RUDOLF CARNAP

FERNANDO HUESCA RAMÓN
Benemérita Universidad Autónoma de Puebla

1 Introduction

The figure of Rudolf Carnap is well known in the field of the history of philosophy, both for his contributions to the formation of the so-called "Vienna Circle" (for instance, through the joint publication of the pamphlet *The Scientific Conception of the World*[1]), or for his polemical attack against Heidegger[2] in the famous *The Elimination of Metaphysics Through Logical Analysis of Language*[3], or for the influential critique that Quine directed against the distinction between "analytic and synthetic"[4] (fully upheld and developed to admirable technical heights by Carnap), as well as against the distinction between the scientific and the mythological, common sense and science, etc., which Carnap defended throughout his philosophical career.[5] Indeed, in the specific area of the historiography of philosophy regarding the reception of the Vienna Circle's thought, Carnap is positioned within the spectrum of *analytic empiricists* (Carnap himself defended the term "logical positivism").[6] to characterize his unique position in terms of philosophy of language and science), and thus, in a dimension opposed methodologically in the philosophy of science, or the philosophical analysis of scientific language, to that of sociological historicism;[7] in a dimension of unnuanced examination, this could even lead to the "popular

---

[1] See [11].
[2] See [9].
[3] See [8].
[4] See [16].
[5] See Überwindung for the distinction between the mythological and the scientific, see [4], as well as for the distinction between natural and artificial language, and between the material mode of speech and the formal mode of speech.
[6] [4, p. 32].
[7] [15, p. 153].

myth" of the rivalry between Carnap and Kuhn, which in turn leads to a one-sided conception that there is an irreconcilable enmity between logical syntax and historicist analysis. Significantly, Carnap himself advocates for a "peaceful coexistence"[8] in his response to Strawson in the 1963 Schilpp Volume, between the motivations for reflection on language in *constructivism* and *naturalism*.

Carnap serves as a meeting point for various trends of thought within the history of philosophy: Leibniz, Hume, Kant, Husserl, Frege, Russell and Whitehead, the Wittgenstein of the Tractatus, Gödel, and Tarski all find in the philosopher and logician from Ronsdorf a place of recognition and exposition in terms of issues related to the relationship between philosophy, logic, and language. Moreover, this is framed in terms of the desirable (for its own sake, we might even venture to add, in the inspiration of the ethical notes in Carnap's texts[9]) of intersubjective communication between human beings[10]; significantly, the older Carnap applauds in his "Intellectual Autobiography" the constructivist motives of Leibniz (even comparing the "characteristica universalis" of the eminent philosopher from Leipzig with the "logical symbolism" or *Begriffsschrift* of Frege), as well as the efforts of the author of the *Monadology* regarding a universal language for "international communication."[11]

Thus, the accusation of dogmatism leveled by Quine (and others[12]) towards Carnap and "modern empiricism," due to the defense of the validity of the analytic/synthetic distinction and "reductionism"[13] The notion of immediate or sensible experience would deserve to be thoroughly nuanced in light of Carnap's own indications regarding the problems of Wittgenstein's conceptions in the Tractatus and the now-called "semantic turn"[14] pointed out in Carnap's *Autobiography* in an open and concise acknowledgment of Tarski for being the first to develop a method for obtaining "adequate definitions of the concept

---

[8] [6, p. 939].

[9] See [5, p. 82].

[10] On the "intersubjective" in a language: "If the statement (Satz) belongs to an intersubjective language (intersubjektiven Sprache) and is to be verified (verifiziert), not only by me but by different subjects S1, S2, etc., then each subject S1 must translate the statement p into a statement p_i within their own phenomenal language (phänomenalen Sprache)." [3, p. 9].

[11] [5, p. 69].

[12] See [18, p. 516]; also [14, pp. 21-38].

[13] [16, p. 20].

[14] [1, p. viii].

of truth."[15] The shift towards confirmationism[16] and semantics[17] pointed out by the German logician himself calls for the expansion of the factual sense[18] motives apparent in the Tractatus, to include all sorts of ways of speaking (such as everyday, non-technical scientific, and formal-technical), fully meaningful in a cognitive and intersubjective sense, in a way that, to this day, invites thought on a Carnapian application of the "logical analysis of language" not so much aimed at dismissing one by one the pseudo-statements of metaphysics (a possible topic of Überwindung), but rather seeking to establish some form of connection or relevance within the everyday and sensibly verifiable spectrum by anyone in some way, of a statement by a journalist, a classical metaphysician, or a philosopher of language versed in some empirical science. Even a consideration of the history of metaphysics in terms of a history of "problems of logic and semantics"[19] is both considered and applied by Carnap in matters concerning the definition of the analytic and the synthetic, the construction of universal languages, and even the different functions (namely, primarily the cognitive and expressive) of language. Perhaps in light of these elements, it should not come as a surprise that even an unsuspected Carnap-Heidegger link is considered by the old Lukács[20]; indeed, both the strictly logical-instrumental and the expressive-experiential aspects find a place in the thought of the German logical positivist.

Thus, the topic we present here is the explication of the relationship between philosophy and language as formulated by Carnap throughout his philosophical life (taking into account his autobiographical notes on The Vienna Circle, The Liberalization of Empiricism, Semantics, Language Planning, etc.). We explore the thesis put forward by Sarkar in 2006 that "by the end of Carnap's life, 'physicalism' meant nothing more than the adoption of a non-solipsistic lan-

---

[15][5, p. 59].

[16]"I proposed to speak of confirmability instead of verifiability. A sentence is considered confirmable if observation sentences can contribute positively or negatively to its confirmation." [5, p. 58].]

[17]"An important step in the development of language analysis consisted in the complementing of syntax with semantics, that is, the theory of the concepts of meaning and truth." [5, p. 59].

[18]"The proposition (Satz) is the description of a state of affairs (Sachverhaltes)." [19, p. 54]; "Tautology and contradiction are not images of reality. They do not represent any possible state of affairs." [19, p. 86].

[19][2, p. 932].

[20]"It seems to us, on the other hand, illustrative that Carnap and Heidegger not only influence present-day thought as opposite extremes, but are extremes of trends that socially stem from the same source, and for this reason, have much in common in their theoretical foundations, and precisely complement each other in such polarity." [13, p. 372].

guage—that is, one in which intersubjectivity is possible."[21] Drawing from texts such as The Overcoming of Metaphysics, The Physicalist Language as the Universal Language of Science, The Logical Syntax of Language, Philosophy and Logical Syntax, Meaning and Necessity, Autobiography, the Schilpp volume, as well as various primary and secondary biographical sources—including unpublished materials and correspondence—we aim to outline a global reconstruction of Rudolf Carnap's philosophy of language.

2 Logical analysis of language

This method of logical analysis (*Methode der logischen Analyse*) is what essentially distinguishes the new empiricism and positivism from the earlier ones, which were more biologically and psychologically oriented. If someone asserts, "There is no God," "The foundation of the world is the unconscious," or "There is an entelechy as a guiding principle in living beings," we do not say, "What you say is false." Instead, we ask: "What do you mean by your statements (*Aussagen*)?" In this way, it becomes clear that there is a distinct boundary between two types of statements. One type includes statements made in empirical science. Their meaning (*Sinn*) can be determined through logical analysis; that is, by referring to extremely simple statements about what is empirically given. The other type includes statements like those we just mentioned, which turn out to be entirely meaningless (*bedeutungsleer*) if taken in the way the metaphysician intends. Often, they can be reinterpreted (*umdeuten*) into empirical statements; however, in doing so, they lose their emotional content (*Gefühlsgehalt*), which is often the most essential aspect for the metaphysician.[22]

This programmatic statement from *The Scientific Conception of the World*, and thus from logical positivism in its early stages as a movement or school, accounts for two of the central motives for reflection and action of the Vienna Circle, namely, the reduction of meaningful statements to empirical statements, as they would be found *par excellence* in the empirical sciences or in experiences of the "given," and whose truth or falsity can be established from "verifiable singular results (*beweisbare Einzelergebnisse*)"[23]; and the motive of the appli-

---

[21][17, p. 83].
[22][11, p. 306].
[23][11, p. 305].

cability of "modern symbolic logic (*modernen symbolischen Logik*) or logistics (*Logistik*)"[24] for the organization of scientific discourse and the critique of traditional metaphysics. Indeed, the bet of the first program (possibly the only one as such, before Hahn's death and the separation of the paths of Neurath and Carnap, the three signatories of *The Scientific Conception*) of the Vienna Circle was primarily framed in terms of verificationism, in the manner of the *Tractatus*, concerning meaning, strictly speaking, and constructivism, in the manner of Frege, regarding a universal notation for the construction of statements. Precisely, we must show how, at least in the case of Carnap, it is possible to trace nuances or grounds for further specification regarding these two topics, in such a way that the accusations of dogmatism from Quine and others could be reconsidered in light of greater bibliographical evidence than that available to the sharp-tongued North American analytic philosopher around 1951.

For example, regarding the critical distance from Wittgenstein of the *Tractatus*, consider the following excerpt from *Philosophy and Logical Syntax*:

> I, along with my friends in the Vienna Circle, owe much to Wittgenstein, especially regarding the analysis of metaphysics. But on the point just mentioned [the Wittgensteinian thesis that says, "Whereof one cannot speak, thereof one must be silent."], I cannot agree with him. Firstly, it seems to me that he is inconsistent in what he does. He tells us that philosophical statements (philosophical propositions) cannot be established and therefore should not be spoken of, but rather one should remain silent; and yet, instead of remaining silent, he writes an entire philosophical book. Secondly, I do not agree with his assertion that all of his statements are as meaningless (without sense) as metaphysical statements. My opinion is that a great number of his propositions (unfortunately not all of them) do in fact have meaning, and the same is true for all the propositions of logical analysis. [4, p. 38]

Indeed, the Carnap who writes a didactic presentation of the more technical development of *The Logical Syntax of Language*[25], focusing on a "formal theory" of an artificial language (with, of course, considerations of natural languages), especially regarding its construction and application to the testing of empirical statements, already in 1934 understands that it is not enough to only speak of "meaning" in relation to the testing of the existence or non-existence

---
[24] [11, p. 308].
[25] [7].

of certain empirical states of affairs, but rather in a much broader relationship that involves the establishment of two orders of validity for the statements inferred from a given scientific language, namely, one logical and one empirical-extralogical. Carnap calls these two orders of validity "L-consequence" and "P-consequence"[26], respectively. For this reason, one should not demand that philosophy and philosophers remain silent on matters unrelated to the assessment of empirical states of affairs, but rather develop a full formal theory of language. This theory would consider, among other things, a logic of construction and derivation (*formation rules* and *transformation rules* in Carnap's terminology) of statements, based on rules, axioms, primitive terms, etc., which, strictly speaking, would have to be expressed in a metalanguage.

Precisely, an important—if not entirely decisive—turn in Carnap's intellectual evolution consists in the inclusion of semantics, understood in terms of a "theory of the concepts of meaning and truth"[27] as a complementary part of a "theory of the forms of expressions of a language,"[28] or syntax. Regarding the revolutionary nature of Tarski's conception on this subject, Carnap states the following:

> When Tarski first told me that he had constructed a definition of truth, I assumed he meant a syntactic definition of logical truth or provability. I was surprised when he said he meant "truth" in the usual sense, including factual truth. Since I was thinking only about problems of a syntactic metalanguage, I wondered how it was possible to establish the truth condition of a simple statement like "this table is black." Tarski replied: "It is simple: the statement 'this table is black' is true if and only if this table is black."[29]

Thus, the redirection of the age-old question of truth toward the consideration of a "metalanguage that contains the statements of the object language or translations of them"[30] opens the path for Carnap's mature perspective (which we might also dare to consider definitive, based on his *Autobiography*) on the possibility of viewing science as a calculus, or a special artificial language. This language would include both a *theoretical language* and an *observation lan-*

---

[26] [4, p. 52].
[27] [5, p. 59].
[28] [5, p. 53].
[29] [5, p. 60].
[30] [5, p. 60].

guage[31], and, by incorporating elements of probability[32], could serve humanity in the *practical applications*[33] of everyday life.

Thus, Carnap's intellectual-biographical development moves from an initial verificationist enthusiasm within the Vienna Circle to a conception of science that, by incorporating both the logical and the historical, is not entirely distant from the balanced perspectives of the older Hempel[34] and Kuhn[35]—balanced in the sense that they precisely consider the issue of the relevance and validity of logic and history in scientific activity. A decisive part of this development is the inclusion of syntax and semantics within a comprehensive program of philosophy of language and philosophy of science. Within this framework, one can find an admirable and fully applicable conception of natural and artificial languages concerning the syntactic construction of statements and their possible factual content. This, in turn, allows for a rigorous discussion of both logical sense and cognitive-empirical sense. From its very origins in the Vienna Circle, this approach has been structurally motivated by a consideration of the expressive and even motivational nature of language.

Finally, all these aspects of exposition lead to Carnap's mature thesis that "one of the most important advantages of the physicalistic language is its intersubjectivity, that is, the fact that the events described in this language are, in principle, observable by all users of the language."[36] Thus, the Carnapian physicalist stance, which can be philosophically characterized, following the German philosopher himself, as a "methodological materialism (methodischer Materialismus),"[37] can be examined in its scope through a bibliographical update that takes into account a broad study sample, allowing for a comprehensive overview of the language theory of the author of *The Logical Syntax of Language*. Regarding other spheres of reflection, such as ethics, the themes surrounding intersubjectivity can be considered or applied within an ideological program that Carnap himself called "scientific humanism."[38]

---

[31] [5, p. 77].
[32] [5, p. 70].
[33] [5, p. 70].
[34] [10, p. 350].
[35] [10, p. 421].
[36] [5, p. 30].
[37] [3, p. 31].
[38] [5, p. 82].

3 Conclusion

The question of the nature and genesis of Carnap's philosophy of language and his commitment to physicalism leads us to pose some follow-up questions, such as: What is the relationship between Carnap's mature theoretical perspective—after the "semantic turn"—and the original project of the Vienna Circle, as well as its debt to Wittgenstein's *Tractatus*? How significant was Carnap's engagement with Tarski's semantic conception of truth for the development of his philosophy of language? What are the fundamental theoretical elements that Carnap invokes regarding the syntax and semantics of language? What concrete applications can the syntactic-semantic method of linguistic analysis developed by Carnap lead to? And finally, what is the relevance of the question of physicalism for our present?

The idea of a "serious crisis" in philosophy in the 21st century, pointed out by Leitgeb[39] in 2009—precisely due to the neglect of Carnap's reflective motives linking logic, philosophy, and science—calls for a re-examination of the philosophical reflections of the author of *The Overcoming* regarding physicalism and intersubjectivity, in order to advocate for a renewed relationship between philosophy, science, and society in the present. Carnap's philosophical-ethical thought, precisely in its breadth, rigor, and consistency, offers at the very least alternative elements of reflection for this contemporary landscape. The social commitment of philosophy and science is, in fact, one of the central motives of the "scientific humanism" program championed by the mature Carnap.

References

[1] BEANEY, M. Foreword. In *Wagner, Pierre, Carnap's Logical syntax of Language,*. Great Britain, Palgrave MacMillan, 2009.

[2] BETH, E. Carnap's views on the advantages of constructed systems over natural languages in the philosophy of science. In *Schilpp, Paul Arthur (ed.): The Philosophy of Rudolf Carnap*. Illinois, Open Court, 1963.

[3] CARNAP, R. Die physikalische sprache als universalsprache der wissenschaft (item 22) 1930-1932, box 110a, folder 3d. In *Rudolf Carnap, Rudolf Carnap Papers, 1905-1970*. Special Collectons Department, University of Pittsburgh, 1930.

[4] CARNAP, R. *Philosophy and Logical Syntax*. London, Kegan Paul, Trench, Trübner and Co. Ltd., 1935.

[5] CARNAP, R. Intellectual autobiography. In *Schilpp, Paul Arthur (ed.): The Philosophy of Rudolf Carnap*. Illinois, Open Court, 1963.

---

[39][12, p. 163].

[6] CARNAP, R. P.f. strawson on linguistic naturalism. In *Schilpp, Paul Arthur (ed.): The Philosophy of Rudolf Carnap*. Illinois, Open Court, 1963.
[7] CARNAP, R. *Logische Syntax der Sprache*. 2nd ed., New York, Springer, 1968.
[8] CARNAP, R. Überwindung der metaphysik durch logische analyse der sprache. In *Mormann, Thomas (ed.), Scheinprobleme in der Philosophie und andere metaphysikkritische Schriften*. Hamburg, Felix Meiner, 2004.
[9] FRIEDMAN, M. Carnap, cassirer, and heidegger: The davos disputation and twientieth century philosophy. *European Journal of Philosophy, 10:3, pp. 263-274* (2002).
[10] GRANDY, R. E. Carl gustav hempel. In *Sarkar, Sahotra y Pfeifer, Jessica, The Philosophy of Science, An Encyclopedia*. New York, Routledge, 2006.
[11] HAHN, HANS, N. O., AND CARNAP, R. Wissenschaftliche weltauffassung: Der wiener kreis. In *Verein Ernst Mach, Veröffentlichungen des Vereines Ernst Mach*. Artur Wolf Verlag, Wien, 1929.
[12] LEITGEB, H. Rudolf carnap's the logical structure of the world. *Topoi* (2009).
[13] LUKÁCS, G. *Prolegomena, Zur Ontologie des gesellschaftlichen Seins, 1*. Halbband, Darmstadt und Neuwied, Luchterhand, 1984.
[14] MAYS, W. Carnap on logic and language. *Proceedings of the Aristotelian Society, New Series, Vol. 62, pp. 21-38* (1961).
[15] PINTO DE OLIVEIRA, J. Carnap, kuhn, and revisionism: on the publication of structure in encyclopedia. *Journal for General Philosophy of Science, 38, pp. 147-157* (2007).
[16] QUINE, W. Main trends in recent philosophy: Two dogmas of empirism. *The Philosophical Review, Vol. 60, No. 1, 195, p. 20-43* (1951).
[17] SARKAR, S. Rudolf carnap. In *Sarkar, Sahotra y Pfeifer, Jessica, The Philosophy of Science, An Encyclopedia*. New York, Routledge, 2005.
[18] STRAWSON, P. Carnap's views on constructed systems versus natural languages in analytic philosophy. In *Schilpp, Paul Arthur (ed.): The Philosophy of Rudolf Carnap*. Open Court, 1963.
[19] WITTGEINSTEIN, L. *Tractatus logico-philosophicus*. Madrid, Alianza, 1991.

Argumentation, Images, and the Concept of Information

Francisco Montes
Universidad Nacional Autónoma de México

Fernando Huesca Ramón
Benemérita Universidad Autónoma de Puebla

1 Introduction

Some years ago, I organized an international conference at Benemérita Universidad Autónoma de Puebla, Mexico. I was in a beautiful conference room located inside a historic building. I was preparing the room for the keynote speaker when he approached to me. At first, I thought he was going to say something but instead he tried to make some signals with his hands. I asked him if he needed anything, but he didn't say a word. I was struggling trying to understand what he wanted to communicate with his hand signals. To this day I really don't know why he didn't say anything. After a moment of bewilderment, he quickly decided to grab some paper sheets and tried to place them on the lectern, but they fell off. Somehow, I immediately understood that what he wanted to communicate was that there was something wrong with the lectern, so I replaced it, and the problem was solved.

I assume that many of us have been in a similar situation where someone tries to communicate something without using any spoken word, only with the help of signals, facial expressions, body postures, images or simply by pointing to an object with the finger. The above described situation shows a person that tries to communicate a seemingly simple fact —that there was a problem with the lectern— by using signals or recreating a failed action.

Possibly one of most common ways to communicate something without using words is through the use of images. People use images to communicate ideas, desires, requests, intentions and even arguments. Barceló-Aspeitia [1] emphasizes the role images play in argumentation. He claims that images "can contribute directly and substantially to the communication of the propositions that play the roles of premises and conclusion." [1] Barceló-Aspeitia refers to "external man-made images, like pictures, symbols, icons, diagrams, maps, etc." [1] My aim in this work is to explain how even though images can help to communicate propositions, sometimes they can fail to provide information to support an argument. I will try to describe how this failure can occur and how it can be explained from Dretske's concept of information.

2 Images as an essential part of argumentation

I will start by examining Barceló-Aspeitia's account on the role that images play in argumentation. According to him "that images are commonly used with persuasive ends and in argumentation is an uncontroversial fact." [1] For Barceló-Aspeitia, what is actually relevant is how images "can play a substantial role in argumentation" [1], so his task will be to show that not all images play "a merely illustrative or ornamental role" [1] within an argument. In this respect, Barceló-Aspeitia offers an account that grants images the same status as words or verbal resources in the process of argumentation, which have been considered "the epitome of argument" [5]. In fact, his main interest is "to topple verbal language from its central place in argumentation, i.e., to show that non-linguistic entities like images can play a role in argumentation (...)" [1]. Although he admits that "[s]ometimes images are used to reinforce or decorate texts, adding nothing substantial to their content" [1], in some other cases, images "do play the kind of communicative role that has traditionally been assigned exclusively to words." [1]

The way in which Barceló-Aspeitia seeks to accomplish his task is by considering some examples of what he calls, following Barwise [2], heterogenous arguments, which are defined as "arguments that are not conveyed through a single medium, but instead make use of both verbal and visual resources." [1] Those examples, in accordance with Barceló-Aspeitia, illustrate different situations where people try to communicate arguments using images. Let us explore those examples.

Example 1 explains the case of Alice and Bruce. Alice tries to explain that there are colored objects by simply picking up a red pen and showing it to Bruce while saying: "Red. Right?" By doing that she seemingly proves that there is at least one colored thing. This example describes an instance of what Barceló-Aspeitia calls subsentential speech, that is, words and phrases which, together with context, can provide propositions used in the construction of arguments. Those words and phrases ("Red") are not considered sentences, even so, following Barceló-Aspeitia, they can communicate a proposition. Subsentential speech combined with complete sentences can provide the elements of argumentation, that is, premises and conclusion. This kind of argumentation is called subsentential argumentation, which are "cases of argumentation where premises or conclusion are conveyed sub-sententially." [9]

According to Barceló-Aspeitia, in Example 1, Alice provides "novel and necessary information for the communication of one of her premises, and did so without need of a verbal intermediary. In other words, what Alice conveyed to Bruce by holding up the pen was the pen itself, and not a word." [1] Specifically, Alice combines "verbally conveyed information and ostensively conveyed information (i.e. information conveyed by showing or pointing at things in the environment)" [1] For Barceló-Aspeitia, this example is a case of an argument where verbal and non-verbal elements are involved, and "it is not hard to see that some of them could have been conveyed by other non-verbal means, like images." [1] So images can fulfill the same function as non-verbal elements in argumentation. Example 1 enables Barceló-Aspeitia to consider images as kinds of non-verbal elements.

Example 2 illustrates the case of Carly and Daniel. Carly tries to convince Daniel that a friend, John, danced at a party. To demonstrate this, she uses a picture to indicate Daniel that one of the portrayed dancers is John. Carly goes on to say: "John. Right?" Then Daniel nods and the discussion concludes.

This example shows an argument where a subsentential phrase and an image are involved. The difference from Example 1 is that this time we are dealing with a non-verbal element in the form of an image (a photograph). The image here plays a key role in the argument, without it the communication of the premise is at risk. The photograph alone carries sub-propositional information and in combination with a subsentential phrase ("John"), we obtained a fully propositional premise [1], which is necessary to form the argument.

In this point, Barceló-Aspeitia argues in favor of his heterogenous argument account. He argues against others who hold that images used in heterogenous arguments are verbally reconstructed, and only after that reconstruction they help in the construction of arguments. He claims, instead, that "propositional premises and conclusions can be conveyed with the aid of visual images, without the need of verbal reconstruction" [1]. For him, such verbal reconstruction cannot be performed either by the hearer or by the speaker. Barceló-Aspeitia, following Stainton [9], holds that:

> Arguers can successfully use a subsentential phrase and a picture to

communicate a full proposition in situations even when there is no information to decide among the many available verbalizations. In these cases, no sentence could have been intended (by the speaker) or recovered (by the hearer). [1]

In other words, it is very difficult to determine which sentence corresponds to a certain image. Barceló-Aspeitia adds that "it is not always possible to translate heterogenous arguments into verbal ones, it is very unlikely that this is what happens every time we interpret heterogenous arguments." [1] This means we can "grasp the proposition without verbalizing it." [1] Example 2, therefore, shows the use of an image in the process of argumentation and helps Barceló-Aspeitia to explain why the use of images does not require verbalization.

Let us now examine the Example 3 as written by Barceló-Aspeitia:

Eugene and Fred are driving through town on the same car, looking for a way out into the highway. They have stopped at an intersection. Federika is at the wheel, and Eugene has a map of the city in his hands. She turns on her blinkers to indicate her plans to make a right turn. Eugene tells her no to do it. "Look," he says, pointing at a section of the map showing that the street Federika was about to turn is closed, "we will never get to the highway that way." [1]

This example shows Eugene using a map to communicate to Federika that the street is closed. This act, according to Barceló-Aspeitia, involves premises and conclusion without using sentences. This would be a case where an image plays a direct and substantial role within an argument.

Barceló-Aspeitia considers that "we do not usually say that maps are true or false. At most, we say that they are accurate or inaccurate (...)" [1]. In a footnote, he adds, following Malinas [6], that "[h]owever, we do speak of propositions being true on them [on the maps]. For example, in Eugene and Federika's argument, we may say that it is true that on the map the street is closed (...)". [1] Additionally, he indicates that if the map was wrong, and the street, in fact, not closed, "[t]he information it conveyed about the street was false". [1] We will discuss later in the text how according to Dretske a map can "say" truly or falsely where a thing stands, and how information cannot be in any case false.

Finally, Example 4 describes a situation in which Hanna, a worker from a petting zoo, tries to determine the correct label for a jar that contains certain type of food. The label has a picture of a specific animal. After short inspection of a jar with food, she decides to label that jar with a rabbit picture. Example 4 is a case about the use of images to communicate a conclusion ("this jar contains rabbit food"). Following Barceló-Aspeitia, Example 3 and 4 are cases where images convey propositional premises and conclusions "without the assistance of any linguistic material, either sentential or sub-sentential." [1]

According to Barceló-Aspeitia, these examples show that images contribute directly to the communication of premises or conclusions, which are the basic components of argumentation, and they can do that "without the need of verbalization or verbal reinforcement." [1] He suggests that this contribution "can be either sub-propositional (...) or fully propositional (...)" [1]. It can be sub-propositional when "properties and functions that, properly combined with information conveyed through other means, like words, or available in the context, can yield full propositions (...)" [1]. Meanwhile, it can be fully propositional when there is no need of verbalization at all.

Barceló-Aspeitia presents a reasonable explanation about how images can play an essential role in the argumentation process. He suggests that images are combined with words or verbal expressions and, together with context, can help to form propositions. In some cases, they can directly contribute to the construction of arguments without the need of verbalization. Assuming that images can be used effectively in the argumentation process, how can images help to support the whole argument? How can they provide evidence to support the argument itself? Or, we can ask instead, how can they fail to support an argument?

3 Dretske's concept of information

So far, I have examined Barceló-Aspeitia's account on how images play a key role in the process of argumentation. He uses some examples to explain the use of images within arguments. I will analyze Dretske's concept of information in order to show how such concept can be useful to understand some issues related to the use of images to support arguments.

In his book Knowledge and the Flow of Information [3], Fred Dretske proposes a theory of information that has influenced discussions in the philosophical literature, especially those related to epistemology, philosophy of mind and cognitive science. For him, it is important for philosophy to collaborate with other disciplines in order to improve our understanding of the concept of information. Consequently, he analyzes the concept of information used in communication theory and tries to apply it to a theory of knowledge. The main idea is that information is needed for knowledge. This idea is expressed in the following passage as Dretske refers to the collaboration aforementioned:

If information is really what it takes to know, then it seems reasonable to expect that a more precise account of information will yield a scientifically more creditable theory of knowledge. Maybe —or so we may hope— communication engineers can help philosophers with questions raised by Descartes and Kant. That is one of the motives behind information-based theories of knowledge. [4]

But what is information? Dretske's first approach to the concept of information is directed to its objectivity. Dretske argues that we can think about information as "an objective commodity, something whose generation, transmission, and reception do not require or in any way presuppose interpretive processes." This idea is emphasized when he notes that "[i]n the beginning there was information. The word came later (...)" [3]. And also when he states that "the raw material is information." (1981) Hence information is an objective product, a material related to knowledge that doesn't need an interpreter. This is clearly seen when he claims that "[i]nformation doesn't need conscious beings to exist, but knowledge does. Without life there is no knowledge (because there is nobody to know anything), but there is still information." [4][1] Certainly, by conscious beings, Dretske refers to intelligent organisms provided with cognitive systems, "those resources for interpreting signals, holding beliefs, and acquire knowledge (...). The higher-level accomplishments associated with intelligent life can then be seen as manifestations of progressively more efficient ways of handling and coding information." [3].

In his text titled "Epistemology and information" [4], Dretske offers a different approach to the concept of information. There, he defines information as an "epistemologically important commodity. It is important because it is necessary for knowledge." [4] Here the emphasis is placed on the epistemic aspect of information. We can learn something from information because "information is that commodity capable of yielding knowledge" [3]. How is information provided? For Dretske, information can be provided through signals, that is, states of affairs, from which we can learn something. A signal is the source of information, it contains and carries information: "(...) one learns, or can learn, from a signal (event, condition, or state of affairs), and hence the information carried by that signal (...)" [3]. In the same sense, Dretske notes, "what information a signal carries is what we can learn from it." [3]. These references

---

[1]This shows us the naturalistic and materialistic ideas behind his notion of information, as Dretske himself indicates: "The entire project can be viewed as an exercise in naturalism –or, if you prefer, materialistic metaphysics." [3] In this respect, if both knowledge and information are directly related concepts, Dretske also proposes a "naturalised" theory of knowledge [7].

to signals are summarized in the following passage: "A state of affairs contains information about X to just that extent to which a suitable placed observer could learn something about X by consulting it." [3]

According to Dretske, states of affairs or signals contain or carry information. Examples of these signals are explained when he writes:

> When a scientist tells us that we can use the pupil of the eye as a source of information about another person's feeling or attitudes, that a thunder signature (sound) contains information about the lightning channel that produced it, that the dance of a honeybee contains information as to the whereabouts of the nectar, or that the light from a star carries information about the chemical constitution of that body, the scientist is clearly referring to information as something capable of yielding knowledge. [3]

Another approach to understand signals is by using the notion of natural signs. Dretske considers that "[i]nformation is something closely related to what natural signs and indicators provide. We say that the twenty rings in the tree stump indicate, they signify, that the tree is twenty years old. That is the information (about the age of the tree) the rings carry."[2] [4]

Dretske considers that natural signs are related to some kind of meaning, natural meaning. Following Grice [?], he notes that "the informational kind of meaning, the kind of meaning in which smoke means (indicates, is a sign of) fire" [4] is called natural meaning. With this kind of meaning, natural meaning, if an event, e, means (indicates, is a sign) that so-and-so exists, then so-and-so must exist." [4] Therefore, as Dretske adds, "[n]atural meaning is what indicators indicate. It is what natural signs are signs of. Natural meaning is information. It has to be true." [4] This excerpt is essential because this is how Dretske tries to associate information to natural meaning, and at the same time he tries to distinguish it from non-natural meaning. Natural meaning is provided through natural signs, signals that carry information about some state of affairs. In this respect natural meaning has to be true. Non-natural meaning is associated to linguistic signals, these signals have a meaning, a conventional meaning [3]

---

[2]In his theory of signs, Charles S. Peirce proposes a kind of sign that resembles the properties of the natural signs just described, and at the same time, he has a similar idea about the interpreter. For Peirce, the index is "physically connected with its object; they make an organic pair. But the interpreting mind has nothing to do with this connection, except remarking it, after it is established." [8] According to Peirce, "[a] weathercock indicates the direction of the wind." [8] The example of the weathercock fulfills the requisites for a Dretskean natural sign.

that we "assign, create or invent (...)" [4]. So, this kind of meaning "need not be true" [4] This is not to say that linguistic signs or symbols cannot carry information, in fact, they can, but only to the extent that their meaning comes from some state of affairs:

> [w]e convey information by using signs that have a meaning corresponding to the information we wish to convey. But this practice should not lead one to confuse the meaning of a symbol with the information, or amount of information, carried by the symbol. [3]

Dretske indicates that "[a]ccording to this usage, then, signals may have a meaning but they carry information. What information a signal carries is what it is capable of "telling" us, telling us truly, about another state of affairs." [3] For instance, Dretske points out, "[i]f nothing you are told about the trains is true, you haven't been given information about trains. At best, you have been given misinformation." [3] I will return to the idea of misinformation shortly. For now, according to Dretske, it is important to understand that if you do not receive any information, you do not receive knowledge at all. Information is an objective fact that does not depend of any interpreter, and is a "useful" [4] and "valuable" [4] commodity, objectively and epistemically relevant. Information must be true.[3]

4 How can images fail to support arguments?

I summarize here what I have explained so far about Barceló-Aspeitia's account on how images play a key role in argumentation, and Dretske's concept of information.

Barceló-Aspeitia argues that images can contribute substantially and directly to the communication of premises and conclusion in argumentation. For him, images provide novel and necessary information to communicate arguments. They can contribute to arguments when they are properly combined with context, words or verbal resources in the construction of propositions. This contribution is called sub-propositional (Example 1). When images contribute directly to arguments, without the assistance of any linguistic or verbal resource, the contribution is considered fully propositional (Example 2 and Example 3).

---

[3]In this respect, if a map is wrong it cannot conveyed any information at all. We could not say that a map was wrong because it conveyed "false information" about some location. At best we can consider the map useful for the construction of a proposition, although it did not conveyed information.

Dretske claims that information is an objective commodity from which we, as conscious beings, can know or learn something. This information can be accesible to us through signals or natural signs, that have a natural meaning that must be true. We can convey information by using linguist signs or symbols whose meaning "need not to be true" [3]. These signs can carry information only when they are capable of yielding knowledge, that is, when they indicate some state of affairs. Now let us analyze Barceló-Aspeitia examples in terms of Dretske's concept of information.

In Example 1, Alice shows a red pen to Bruce and at the same time uses verbal expressions to communicate her argument. This is a case of ostensive information, that is, "information conveyed by showing or pointing things in the environment" [1]. Following Dretske, we can understand the action of showing a red pen to someone as an event or signal where information is directly perceived.[4] This signal is combined with a verbal phrase in order to communicate a proposition. This case can also be explained with Dretske's distinction between analog and digital information.

According to Dretske, "every signal carries information in both analog and digital form" [3]. For example, following Dretske, if we want to communicate that a cup has coffee in it, we can verbally express that "piece of information" [3]. If we do so "this (acoustic) signal carries the information that the cup has coffee in it in digital form. No more specific information is supplied about the cup (or the coffee) than that there is some coffee in the cup." [3] But, if we decide to provide an image, "a photograph of the scene and show you the picture, the information that the cup has coffee in it is conveyed in analog form." [3] It is analog because the picture carries additional information: "The picture tells you that there is coffee in the cup by telling you, roughly, how much coffee is in the cup, the shape, size, and color of the cup, and so on." [3] If we attempt to express digitally all the information a picture carries, the result will be an "enormously complex sentence, a sentence that describes every detail of the situation about which the picture carries information." [3] Thus, an image is a signal that carries information in analog form and any description from that image will be consider information in digital form. In Example 1 there is no image involved. Since Alice shows a red pen, she is using information available from the environment, information that we can directly perceive, as information in analog form. At the same time, she uses a verbal phrase, an acoustic signal that provides information in digital form. Let's now consider Example 2.

---

[4]Likewise, Mares states: "A signal is an event, most commonly (in Dretske's view) an event that is perceived." (2024)

Example 2 illustrates a case where an image is used to form an argument. Carly uses a photograph to show to Daniel that John was dancing at a party. Suppose that John was really at the party and someone took a picture of him dancing. That picture is the same that Carly showed to Daniel. The image carries the information that corresponds to an event that actually occurred. For Dretske, in this case, the photograph is truly telling Daniel something about one specific state of affairs. When he sees the photograph, he is receiving the information that John was dancing at a party. There is no misinformation or "false information" at all. Furthermore, according to Dretske, "false information and mis-information are not kinds of information —any more than decoy ducks and rubber ducks are kinds of ducks. And to speak of certain information as being reliable is to speak redundantly." [3] Moreover, as we have previously explained, "[i]nformation is what is capable of yielding knowledge, and since knowledge requires truth, information requires it also." [3] The photograph shows an event that occurred, it provides a large quantity of information (information in analog form), one piece of such information is that "John was dancing at the party" (information in digital form), as Carly herself expressed. Daniel sees John in the photograph and all that he can do is to accept Carly's claim because, in fact, it is true. This would be the ideal case.[5]

Consider, by contrast, the following problem. Suppose that the photograph does not portray John but a man that physically resembles John. The photograph still carries information but not the one that indicates that "John was dancing at the party". Carly has reasons to think that the person dancing is John because she has seen him at parties before. But that day John did not assist to the party. She uses that photograph to show Daniel that John was dancing but the truth is that John was not there at the time. In this case she is showing a photograph that carries information but not the one that she describes. She is claiming something that did not happen. The image does not carry the information that John was dancing. It carries information but the one that Carly claims. Here Carly ignores that the man in the photograph is not John. The image is not carrying the information that John was at the party so what she is claiming about the image is false. We can say here that Carly fails to provide a good resource to support her argument.

But the image itself can fail to provide the required information about some state of affairs. Device technical limitations, human error when operating the device, or a sudden change in the environment can affect a photographic record.

---

[5]The ideal case can be expressed by the following statement: "In Dretske's view, all real information is true. If the event that A carries information that B and the event that A really occurs, then B is true." [7]

Here the image (and the user of that image) will fail to provide information about something. Following Example 2, it is possible that the quality of the photograph is not good enough. Actually, John was there, and Carly saw him at the party, but the photograph is not good to prove it. In Dretske's terms, the event really occurs. Daniel has reasons to doubt about Carly's claim. The photograph here (and the use of it) is not helpful and fails to support an argument. The photograph is not providing sufficient information about John. It carries information about John but not as much that is needed to recognize him. Dretske examine this idea when he explains the problem about the content of signals, that is, how much information a signal can carry. That problem alone regarding the informational content of a signal demands a full text. But, for the example above, it would be useful to indicate some remarks. Dretske states that:

> [I]f a signal is to carry the information that s is F (where s denotes some item at the source), then the amount of information that the signal carries about s must be equal to the amount of information generated by s's being F. If s's being F generates 3 bits of information, no signal that carries only 2 bits of information about s can possibly carry the information that s is F. [3]

In this way, a photograph must carry enough information about the source, otherwise it will not carry the appropriate information about that source, even if the event did occur: "The signal carries some information about the source, but not enough to carry the message that [X] (...) occurred." [3] Additionally, Dretske says:

> What communication theory (...) tells us is that for the communication of content, for the transmission of a message, not just any amount of information will do. One needs all the information associated with that content. If an event's occurrence generates X bits of information, then a signal must carry at least X bits of information if it is to bear the information that that event occurred. Anything short of X bits is too little to support this message.

Carly's photograph must be good enough to be used for her argument. If it clearly shows John, and John was at the party, the image carries the information about an event that occurred, and that image will support her argument. If she erroneously takes another man as John, she is claiming something that is not true, the image carries information that is not related to the statement

"John was dancing at the party". If she uses a photograph that does not show clearly John, though he was in fact at the party, the image fails to support her argument. It can be possible that she and the image itself fail to support the argument.

Consider now Example 3, which illustrates the case where Eugene used a map to communicate to Federika that a street is closed. That leads them to conclude that if they take that street they will not get to the highway. They use a map, an image that represents the locations and the conditions of the terrain of a certain area. Or at least, that is what a map purports to show. In this respect, Dretske asks the following questions: "How is it possible for the colored lines, dots, and areas on the paper to misrepresent certain features of an area's geography? What enables the map to say, truly or falsely, as the case may be, that there is a park here, a lake there?" [3]. The answer, certainly, is related to the "information-carrying role". [3] For Dretske, cartographical symbols "are used to convey information about the location of streets, parks, and points of interest in the city." [3] Since the symbols are more or less arbitrary, Dretske continues, "their information-carrying capacity must be underwritten by the intentions, integrity, and executive fidelity of the people who make the maps. A crucial link in the flow of information (from the physical terrain to the arrangement of marks on paper) is the map maker himself." [3] Of course, a map can fail to convey information about some location not just by "ignorance, carelessness, or deceit" [3], but by a variety of reasons. For example, it could be the case that locations and conditions of the terrain have changed, and maps of the area are not yet actualized. Nonetheless, the answer is the same, the symbols of the map are not carrying the information they are supposed to carry about some state of affairs, even when such symbols are still indicating, for example, that a lake is in some specific place. Consequently:

> a map can misrepresent the geography of an area only insofar as its elements (the various colored marks) are understood to have a meaning independent of their success in carrying information on any given occasion. A particular configuration of marks can say (mean) that there is a lake in the park without there actually being a lake in the park (without actually carrying this piece of information) (...). [3]

In other words, the symbols and the marks have an "information-carrying function", but they fail to carry information, they are "saying" something about certain area that does not represent an actual state of affairs. A map can misrepresent how things stand, and misrepresentation, according to Dretske, is "meaning without truth". [3][6]

Recalling Barceló-Aspeitia's note about the truth and falsehood of maps, if information is always true, as Dretske indicates, a map that is wrong cannot conveyed "false information". Perhaps it can only "misinform" (see [4]). On the other hand, Barceló-Aspeitia states, following Malinas [6], that "we may say that it is true that on the map the street is closed" [1]. If we follow Dretske instead, we may say that what the map is showing "means" that the street is closed, and if the cartographic symbols do their job and carry that information, we are in the position to say that the map represents truly a certain location.

Finally, Example 4 consists of an image used as a conclusion of an argument. The example illustrates a situation in which a zoo worker inspects food from a jar and decides to label that jar with a picture of a rabbit, indicating that the jar contains rabbit food. This example can be explained similarly as in Example 3. We could say that the image in the jar has some "meaning" (a conventional meaning), and that meaning is "This jar has rabbit food". This will be true if the food in the jar is indeed rabbit food. Now consider this alternative situation. Suppose that the jar with the image depicting a rabbit does not have any food or it contains food for a different animal. The jar labelled with a picture of a rabbit would still has the meaning "The jar has rabbit food", although it is not true. We could say that it is true that the jar has rabbit food only if the jar actually has rabbit food. It could be a case of misrepresentation, that is, a case of meaning without truth, similarly to the map case in Example 3.

5 Conclusion

My purpose was to extend Barceló-Aspeitia's ideas about the use of images in argumentation by considering how images sometimes fail to support arguments, how this failure can occur and how it can be explained from Dretske's concept of information. Barceló-Aspeitia offers a reasonable explanation on the key role images play in the process of argumentation. Images can be combined with verbal phrases to communicate propositions, but they also can be used directly as propositions or conclusions in arguments. The next step is to think how they can be useful to support arguments. Dretske's concept of information can help us to understand how a signal can communicate something about a state of affairs, how they can carry information about a specific event. Because

---

[6]In Dretske's terms, we could say that maps are signals with non-natural meaning, that is, they are signals that "need not be true" and have a conventional meaning.

images can be used as signals, they are able to carry information. At times we fail to use images to communicate something, at other times the image itself can fail to carry the information we need, and when that situation occurs, it is reasonable to think that they can fail to support arguments. If they fail to carry information, following Dretske, they will not be saying anything about the real world. Certainly, we want images to be useful in the process of argumentation, but we want also images to be useful and reliable to support arguments, and they can do so if they are really saying something about the world.

References

[1] BARCELÓ-ASPEITIA, A. Words and images in argumentation. *Argumentation 26, 355-368* (2012).

[2] BARWISE, J. Heterogenous reasoning. in working papers on diagrams and logic. In *Working Papers on Diagrams and Logic, ed. Jon Barwise and Gerard Allwein, 1–13*. Bloomington: Indiana University Logic Group Preprint No. IULG-93-24, 1993.

[3] DRETSKE, F. *Knowledge and the Flow of Information*. Cambridge, MA: MIT Press, 1981.

[4] DRETSKE, F. Epistemology and information. In *In Handbook of the Philosophy of Science. Volume 8: Philosophy of Information, ed. Pieter Adriaans and Johan van Benthem (General Editors: Dov M. Gabbays, Paul Thagard and John Woods)*. Elservier-North Holland, 2008.

[5] GROARKE, L. Towards a pragma-dialectics of visual argument. In *Advances in pragma-dialectics, ed. F.H. van Eemeren, 137–151*. Amsterdam: SicSat, and Newport News: Vale Press, 2002.

[6] MALINAS, G. A semantics for pictures. *Canadian Journal of Philosophy 21: 275–298* (1991).

[7] MARES, E. *Logic and Information*. Cambridge, MA: Cambridge University Press, 2024.

[8] PEIRCE, C. S. *The Essential Peirce. Selected Philosophical Writings. Volume 2 (1893-1913)*. Bloomington and Indianapolis: Indiana University Press, 1988.

[9] STAINTON, R. J. *Words and thoughts: Subsentences, ellipsis and the philosophy of language*. Oxford: Oxford University Press, 2006.

Topology and Peirce's Existential Graphs: The Case of the Möbius Strip

PANIEL REYES CÁRDENAS
Oblate School of Theology & The University of Sheffield

JOSÉ MARTIN CASTRO

1 Introduction

This paper explores the intersection of topology and Charles Sanders Peirce's existential graphs, focusing on the Möbius strip's potential for representing paradoxical concepts like the "round square" as a structure representable in Existential Graphs. The Möbius strip, a non-orientable surface, allows for the coexistence of a proposition and its negation without logical contradiction. This property can be leveraged to depict the "round square," a traditionally paradoxical concept, by simultaneously representing "roundness" and "squareness." In this chapter, I propose two alternatives to the paradoxes raised by the "round square", one related to set theory and the other, a novel answer drawn from Peirce's existential graphs. The paper also highlights Arnold Oostra's contributions to the field of existential graphs. Oostra emphasises the diagrammatic nature of existential graphs, suggesting their potential as a visual language for reasoning and communication. By combining the topological properties of the Möbius strip with the expressive power of existential graphs, we can gain a deeper understanding of paradoxes and their representation in logical systems.

Existential graphs, developed by Charles Sanders Peirce, provide a diagrammatic representation of logical relations. This work explores the intersection of topology and existential graphs, specifically focusing on the Möbius strip and its implications for representing paradoxical concepts like the "round square", I offer[1] two alternatives to approach the conundrum of a "round square" one

---

[1] I must give credit for this solution to the joint work made with my colleague Martín Castro-Manzano, whose mastery of the graphs in relationship with other logical systems of diagrammatic representation helped us propose this alternative.

alternative related to set theory while the other alternative will emerge from Peirce's existential graphs. This will lead us to some topological implications similar to the ones pointed by Oostra, as it will be shown in the last section.

2 Peirce's Existential Graphs: A Diagrammatic Exposition of Logic

Charles Sanders Peirce, the renowned American philosopher and logician, bequeathed to posterity a remarkable system of visual logic known as Existential Graphs (EG). With characteristic ingenuity, Peirce conceived these graphs as "a diagram to illustrate the general course of thought," contending that they captured the essence of logical reasoning in a manner more intuitive and readily grasped than traditional symbolic methods [6, CP 4.394]. Indeed, these graphs transcend mere symbolic representation, constituting a veritable visual language with which to reason and articulate complex ideas [8].

Existential Graphs are organised into a hierarchy of three subsystems, each addressing a distinct level of logical sophistication (see [6, CP 4.433]):

- **Alpha Graphs:** Forming the bedrock of the system, Alpha graphs are concerned with the representation of classical propositional logic. They provide a visual syntax for expressing the fundamental logical connectives – conjunction, disjunction, and negation – and the relationships between propositions. The 'sheet of assertion' serves as the foundational space upon which propositions are inscribed. Negation is denoted by a 'cut', a closed line drawn upon the sheet of assertion, encircling the proposition to be negated. Peirce, in his own words, described the sheet of assertion as "a surface of indefinite extent" and the cut as "a self-returning cut" [6, CP 4.435, 4.438].

- **Beta Graphs:** Ascending to a higher plane of complexity, Beta graphs delve into the domain of first-order logic, encompassing predicates and quantifiers. They furnish the means to represent individuals, their properties, and the relations between them, thus enabling the expression of more intricate logical assertions. The distinctive feature of Beta graphs is the introduction of 'lines of identity', also termed "heavy lines" by Peirce, which serve to denote individuals and the predicates affirmed of them. [6, CP 4.441; 4.450]

- **Gamma Graphs:** Occupying the summit of the hierarchy, Gamma graphs extend the expressive power of the system to embrace modal logic.

They introduce the modalities of possibility and necessity, permitting the representation of modal propositions such as 'possibly true' or 'necessarily false'. To achieve this, Gamma graphs employ an array of graphical devices, including 'tinctures' and 'scrolls', to signify modal operators and higher-order logical constructs. Peirce, in his explorations of Gamma graphs, experimented with diverse methods for representing modality, including the use of coloured areas and broken cuts. [6, CP 4.510-4.572] and [10]

Peirce's development of Existential Graphs arose from his profound investigations into the logic of relatives, which we now recognise as predicate logic. While conventional first-order logic relies upon algebraic notation, Peirce, with his characteristic penchant for diagrammatic representation, sought a more visually perspicuous approach, convinced that such a representation would foster deeper understanding and facilitate logical reasoning [12]. He esteemed Existential Graphs as his "chef d'œuvre", the culmination of his endeavours to construct a genuinely iconic system of logic [6, MS 478]. On this, Arnold Oostra states:

> "Existential graphs grew out of Peirce's seminal studies in the logic of relatives, now known as predicate logic. While current first-order logic essentially retains Peirce's algebraic presentation of quantifiers, he himself sought a diagrammatic version, finding in existential graphs "a diagram to illustrate the general course of thought" and considering them his *chef d'œuvre*." [5, p. 103]

Existential Graphs, with their hierarchical structure and visual clarity, offer a potent instrument for the exploration of logical concepts and their interrelations. They forge a bridge between the abstract realm of logic and the intuitive domain of visual thinking, opening up new vistas for comprehending and articulating complex ideas [10].

3 The Möbius Strip and the Representation of Paradoxes

In *Imagine a Round Square*, Moktefi and Družinina [3] argue that we can employ diagrams in order to represent strange items as round squares. In this contribution, we argue that their proposal, albeit cutting-edge, will not suffice, but we need to understand the relevance of such a diagrammatic challenge. Indeed, Moktefi and Družinina [3] propose using diagrams to represent seemingly impossible objects like "round squares." Building on this idea, we argue that the Möbius strip, a curious object in topology, offers a unique framework for understanding and representing paradoxes.

The Möbius strip possesses only one side and one edge, defying our intuitive notions of boundaries and surfaces. This "non-orientable" nature arises from its construction: a strip with a half-twist, joined at its ends. [1] This challenges the distinction between "inside" and "outside." This non-orientability has profound implications for representing paradoxes—self-contradictory statements or those leading to contradictions. [9] The Möbius strip provides a visual and conceptual model for how such contradictions might coexist without violating logical consistency.

Imagine a sheet of paper representing a logical assertion. On one side, we write "This statement is true," and on the other, "This statement is false." Traditionally, these cannot both be true. But if we transform the paper into a Möbius strip, the opposing statements become linked on a single, continuous surface. (This idea draws inspiration from the liar paradox and its visual representations; see [7]).

This "topological manipulation" allows us to visualise the coexistence of a proposition and its negation without contradiction. As we traverse the Möbius strip, we seamlessly move from affirmation to negation, highlighting the interconnectedness of seemingly opposed concepts. The Möbius strip's capacity to accommodate paradoxes extends beyond simple logical statements. It can also represent more complex paradoxes, like the "round square." In Euclidean geometry, squares and circles are fundamentally incompatible. However, the Möbius strip, with its square-like edge and circular surface, offers a way to reconcile this apparent contradiction.

The Möbius strip, with its unique topological properties, provides a powerful tool for representing and understanding paradoxes. It challenges our conventional understanding of boundaries and surfaces, offering a visual and conceptual framework for reconciling seemingly contradictory concepts. This opens new avenues for exploring logic, language, and the nature of reality.

If we conceive of the edge of a Möbius strip as delineating a square, with its four distinct sides, and the continuous surface as representing a circle, with its seamless curvature, we can perceive how the Möbius strip embodies the paradoxical nature of a "round square." The non-orientable essence of the strip allows for the simultaneous representation of both "roundness" and "squareness" without engendering a contradiction.

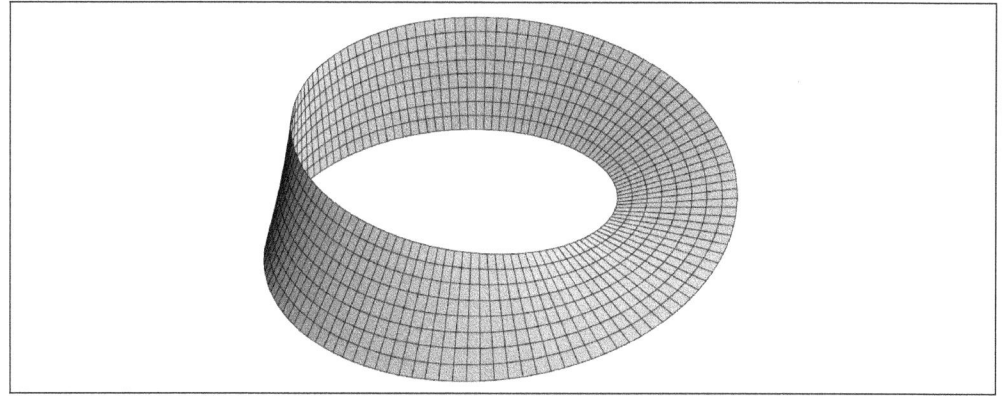

Figure 1: Möbius strip

4 Alternative Representations of the Round Square: Set Theory and Imaginary Numbers

While the Möbius strip provides a compelling topological representation of the round square, it is not altogether clear how set theory could make use of it. This section will show the leverage achieved by set theory on the different mathematical concepts to capture this paradox.

4.1 Set Theory and Venn Diagrams

One approach utilises set theory and Venn diagrams. We can represent the round square by considering the intersection of two sets: the set of all-round objects and the set of all square objects. [11] While this intersection is empty in Euclidean geometry, where roundness and squareness are mutually exclusive, we can conceive of a "round square" as an object belonging to both sets simultaneously. (This relates to the idea of "impossible objects" in cognitive psychology; see [2]). This object might possess characteristics of both roundness and squareness, perhaps exhibiting a continuous curve that also forms a closed, four-sided figure.

4.2 Imaginary Numbers and the Complex Plane

Another approach involves imaginary numbers and their geometric interpretation in the complex plane. [4] We can imagine a "round square" as a figure in the complex plane, where the constraints of Euclidean geometry are relaxed.

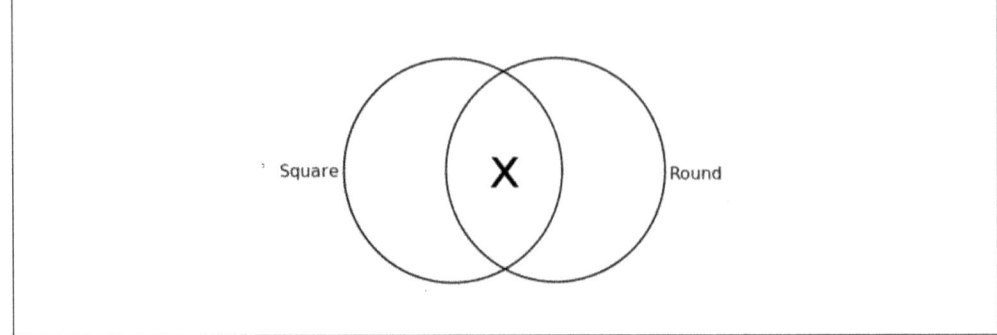

Figure 2: Venn's diagram for a "round square"

This figure might possess both round and square characteristics, perhaps with sides that have imaginary lengths. Unfortunately, this expressivity is not available for traditional set theory, we need a different system (an open-ended one, allowing for further transformations consistent with the rules of the system) of diagrammatic representation that could account for such issues. A sound candidate for this is Peirce's Existential Gamma Graphs.

4.3 The Topological and Existential Graph's Argument for a Local "Round Square"

We agree that diagrams are instruments designed to achieve a specific function within a specific context and that may be manipulated with imagination, but we think that does not imply that diagrams are not "pictorial images" of the objects they stand for. At most, we think that implies that diagrams need not be pictorial images, but it does not follow they cannot be pictorial. So, consider the next argument:

1. Notions and notations are distinct. Notions are concepts, notations are representations of concepts.

2. Our distinction between notions and notations should be relevantly similar across all symbolic fields.

3. The sentence "$x^2 + 1 = 0$" has no real solution. Thus we use an imaginary solution, namely, an imaginary number.

4. But an imaginary number is a number (notion), not the representation of a number (notation).

5. Now, the expression "square circle" has no real image; hence we may have use for an imaginary solution, namely, an imaginary diagram.

6. Hence an imaginary diagram is a diagram, not the representation of a diagram.

Peirce's Existential Graphs (EG) come in a variety of ways: Alpha is the basic system, while Beta and Gamma are extensions of Alpha.1 Alpha is concerned with propositional logic, Beta with first-order logic, and Gamma with modal logic. In Alpha, and hence in the rest of systems, there are three basic elements: the sheet of assertion, the cut, and the literals.

The sheet of assertion (which we represent with a grid in Figure 3a) is a surface in which the two basic features of logic take place: affirmation (represented by the presence of literals within a sheet of assertion, as in Figure 3b), and rejection/negation (represented by a closed line, called "cut," within a sheet of assertion, as in Figure 3c).

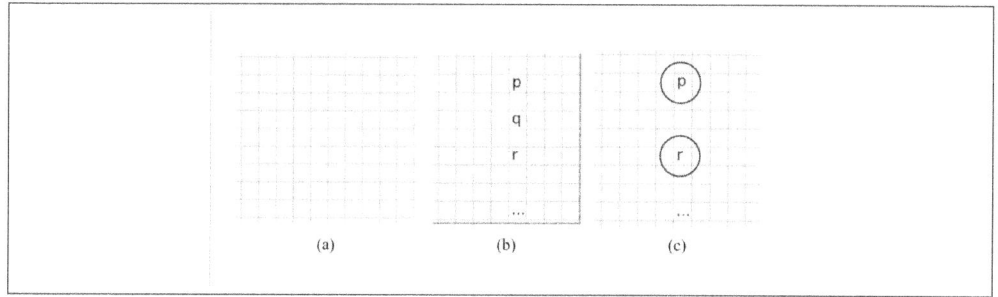

Figure 3: Existential graphs representing a "round square"

Now, let us consider this case:

1. Each face of a sheet of assertion is an orientable surface.

2. If an orientable surface is consistent, it has a model.

3. Suppose p and $\neg p$ are in opposite faces of the same sheet of assertion. If this is the case, each face is an orientable, consistent surface with a model (from 1 and 2).

4. Now, transform the sheet of assertion into a Möbius strip: what we obtain is a single non- orientable sheet of assertion in which both p and $\neg p$, but has a model.

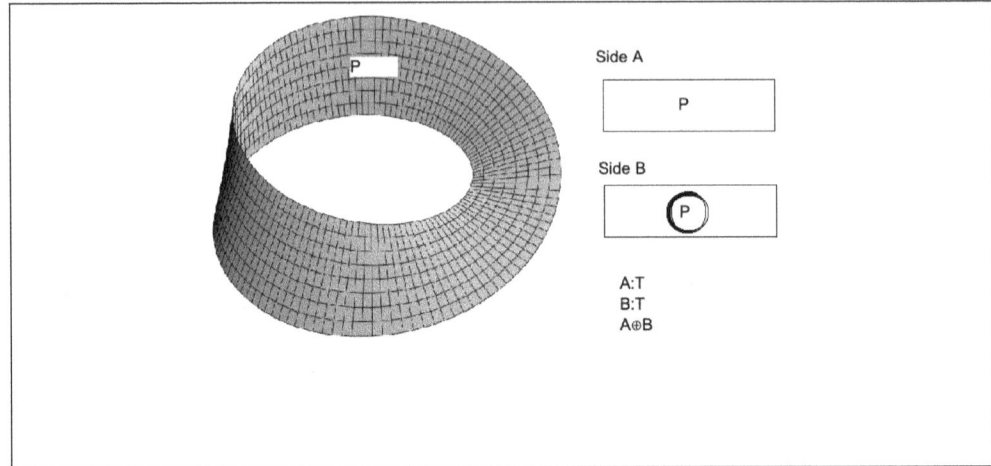

Figure 4: Diagrammatic representation in a Möbius strip as "sheet of assertion"

Thus, we can represent a square circle in a Möbius strip, thus producing a Peircean dialetheia given this situation:

- Non-orientable surfaces <> true contradictions
- Orientable surfaces <> non-true contradictions

Now, the question in order is whether reality is orientable or not.

This analysis, employing the formal framework of Peirce's Existential Graphs and the topological properties of the Möbius strip, yields a significant implication: the possibility of representing a "local" round square, a seemingly paradoxical entity, within a non-orientable logical space. By mapping the conventional sheet of assertion onto a Möbius strip, we construct a unique domain where contradictions, such as the simultaneous affirmation and negation of a proposition (p and ¬p), can be instantiated without violating the law of non-contradiction. This highlights a key advantage of Peirce's system: its capacity to accommodate true contradictions, a feature absent in traditional set-theoretic diagrams constrained by the limitations of Euclidean geometry. The sheet of assertion, as a surface for inscription, provides a more expressive foundation for exploring logical relations, particularly when extended into non-orientable domains.

This raises a crucial question regarding the ontological status of reality itself. If reality, analogous to the Möbius strip, exhibits non-orientable characteristics, then the existence of true contradictions, or *dialetheias*, may not be merely a logical abstraction but rather a reflection of the inherently paradoxical nature of the real. This opens a fertile ground for further investigation into the relationship between topology, logic, and ontology.

Specifically, this framework suggests a potential avenue for resolving the apparent conflict between classical logic, with its strict adherence to the law of non-contradiction, and the existence of paradoxes that seemingly defy this law. By positing a non-orientable logical space, we allow for the possibility of true contradictions without undermining the foundational principles of logical reasoning. This necessitates a re-evaluation of our understanding of logical consistency and the limits of representation, particularly within the context of diagrammatic systems.

5 Arnold Oostra's Advancements

Arnold Oostra's work on existential graphs (EG) has significantly advanced our understanding of their structure, interpretation, and potential applications [5]. He states the inherently diagrammatic nature of EG, arguing that they constitute a visual language for reasoning and communicating complex ideas, echoing Peirce's own vision of EG as "a diagram to illustrate the general course of thought" [5]. This perspective aligns with and strengthens our argument for representing a "local" round square within the non-orientable space of a Möbius strip.

Oostra's analysis clarifies the formal semantics and expressive power of EG, demonstrating their capacity to capture logical relations beyond the limitations of traditional systems. This lends further support to our claim that EG, particularly when mapped onto a Möbius strip, can accommodate true contradictions without compromising logical consistency.

Oostra's contributions provide a valuable theoretical framework for our analysis, reinforcing the validity and significance of representing a "local" round square within the non-orientable space of a Möbius strip using Peirce's Existential Graphs. His work highlights the expressive power and cognitive potential of EG, further strengthening our argument for their unique capacity to accommodate and illuminate paradoxical concepts. Arnold Oostra masterfully extends Peirce's Alpha graphs to encompass intuitionistic logic. He provides a concise historical overview of constructivism and meticulously outlines the necessary modifications, ingeniously utilising Peirce's "scroll" to represent intuitionistic implication and disjunction. Oostra presents transformation rules for the propositional fragment, along with illustrative examples and key metatheorems. Furthermore, he outlines extensions to Beta and Gamma graphs for intuitionistic predicate and modal logics, respectively, and discusses connections to super-intuitionistic logics, opening exciting avenues for philosophical and mathematical exploration. Oostra offers a visionary proposal for a higher-dimensional generalisation of Alpha graphs, underscoring the power of diagrams as abstract tools for representing formal theories, this will be a good framework for projects such as the one presented here: the non-Euclidean projection of the sheet of assertion for Existential Graphs.

6 Conclusion

The Möbius strip and existential graphs offer a unique perspective on representing paradoxical concepts. The non-orientable nature of the Möbius strip allows for the coexistence of contradictory properties, while existential graphs provide a diagrammatic framework for expressing logical relations. By combining these two approaches, we can gain a deeper understanding of the nature of paradoxes and their representation in logical systems.

References

[1] ADAMS, C. *The Knot Book: An Elementary Introduction to the Mathematical Theory of Knots*. American Mathematical Society, 2010.

[2] CHATTERJEE, A. *The Aesthetic Brain: How We Evolved to Desire Beauty and Enjoy Art*. Oxford University Press, 2016.

[3] MOKTEFI, A., AND DRUŽININA, J. Imagine a round square. *Logique et Analyse, 63(251), 281-304* (2020).

[4] NEEDHAM, T. *Visual Complex Analysis*. Oxford University Press, 1997.

[5] OOSTRA, A. Peirce's existential graphs. In *The Routledge Handbook of Logic and Diagrams (pp. 103-121)*. Routledge, 2023.

[6] PEIRCE, C. S. *Collected Papers of Charles Sanders Peirce (Vols. 1-8)*. (C. Hartshorne and P. Weiss, Eds.). Harvard University Press, 1931-1958.

[7] PENROSE, L. S., AND PENROSE, R. Impossible objects: A special type of visual illusion. *British Journal of Psychology, 49(1), 31-33* (1958).

[8] ROBERTS, D. D. *The Existential Graphs of Charles S. Peirce*. Mouton, 1973.

[9] SAINSBURY, R. M. *Paradoxes*. Cambridge University Press, 2009.

[10] SHIN, S.-J. *The Iconic Logic of Peirce's Graphs*. MIT Press, 2002.

[11] VENN, J. On the diagrammatic and mechanical representation of propositions and reasonings. *The London, Edinburgh, and Dublin Philosophical Magazine and Journal of Science, 9(59), 1-18* (1880).

[12] ZEMAN, J. J. *The Graphical Logic of C.S. Peirce*. PhD thesis, Doctoral dissertation, University of Chicago, 1964.

AN AMBIVALENT INTERPRETATION OF THE *DICTUM DE OMNI* IN *PRIOR ANALYTICS* B, 22, 68A16-68A25

LUCA GILI
University of Chieti and Pescara & University of Vilnius

1 Introduction

In recent years, scholars have been discussing whether Aristotle subscribed to the so-called "orthodox" or "heterodox" reading of the *dictum de omni et de nullo*. The label is due to Jonathan Barnes, who maintained in his John Locke Lectures (published in [1]) that Aristotle had an "orthodox interpretation" of the *dictum*, whereas the late Michael Frede, then in the assistance, proposed what Barnes later called the "heterodox reading".[1] There has been a heated debate as to what is the right interpretation of Aristotle's intention, and this paper is designed to reassess the *dictum* and to show that Aristotle might have had in mind an ambivalent interpretation of the *dictum*, that allows for both the orthodox and the heterodox version, depending on the context. But what are these readings? And what is the *dictum de omni* in Aristotle's works?

In the opening pages of his *Prior Analytics*, Aristotle provides a series of definitions and elucidations that will serve as the foundation for the whole syllogistic system, that he later outlines in chapters 2-22 of the first book of the treatise. Aristotle writes[2]:

---

[1] Cf. [18] who mentions the debate between J. Barnes and M. Frede while reviewing [1].

[2] In the following quotation, I reproduce the text printed in the edition by I. Bekker and T. Waitz, who both preferred the reading attested by the manuscript Urbinas gr. 35 (*siglum*: A), μηδὲν ἦ λαβεῖν τῶν τοῦ ὑποκειμένου (I modify Barnes' translation accordingly). W. D. Ross, on the other hand, preferred the reading μηδὲν ἦ λαβεῖν τοῦ ὑποκειμένου. Malink [11] argues in favor of the lectio selected by Bekker and Waitz on philosophical grounds, and I also think that it should be preferred, also on the basis of philological reasons (the consensus of the other manuscripts appears to be an attempt at simplifying the A variant, which is unlikely to be a *lectio singularis eliminanda*).

That one term should be in another as in a whole is the same as for the other to be predicated of all of the first. And we say that one term is predicated of all of another, whenever nothing can be found among the things said of the subject of which the other term cannot be asserted; 'to be predicated of none' must be understood in the same way. (Aristotle, *Prior Analytics*, A, 1, 24b27-24b30, Oxford translation revised by J. Barnes, slightly modified) [1].

Τὸ δὲ ἐν ὅλῳ εἶναι ἕτερον ἑτέρῳ καὶ τὸ κατὰ παντὸς κατηγορεῖσθαι θατέρου θάτερον ταὐτόν ἐστιν. λέγομεν δὲ τὸ κατὰ παντὸς κατηγορεῖσθαι ὅταν μηδὲν ᾖ λαβεῖν τῶν τοῦ ὑποκειμένου καθ' οὗ θάτερον οὐ λεχθήσεται· καὶ τὸ κατὰ μηδενὸς ὡσαύτως.

This passage expands what is said of all, i.e., literally, the "*dici de omni*" (in Latin translation). Among logicians, it has become customary to speak of the "*dictum de omni*" when referring to this paragraph of Aristotle's *Prior Analytics*, though some purists use "*dici*" instead of "*dictum*" to refer unambiguously to Aristotle and not to later reworking of the doctrine by other logicians.

This text allows for two interpretations.

A) Orthodox reading
Aristotle might be arguing that A is said of all B, if and only if, for every individual x, if Bx, then also Ax is the case.

B) Heterodox reading
According to this reading, we do not analyze the phrase in terms of predicate letters and individual variables, but we say that A is said of all B, if and only if, for any predicate C, if B is C, then A will be C as well. This reading is sometimes referred to as "mereological" in the literature because it is tantamount to stating that, if B is a part of A, then, if C is part of B, C will be a part of A as well. These relations between predicate terms do not mention any of the individuals included in the classes.

This distinction could probably be better expressed in Aristotle's own jargon in the following way.

A*) Orthodox reading
A is said of all B iff all the primary substances (or the individual accidents) of

which B is predicated are also A.

B*) Heterodox reading
A is said of all B iff for any subspecies of B (say: C), it will also be true that A is said of C.

There are merits for either reading of Aristotle's text.[3] If one where to assume the heterodox reading, Aristotle could not be *defining* the predicative relation "said of all", because the same relation would occur in the *definiendum* (e.g., "A is said of all B") and in the *definiens* (e.g., "for all C, if B *is said of all C, then A is said of all C*"). On the other hand, the supporter of the orthodox reading could say that their interpretation avoids this problem, because the predicative relation between two predicate letters (e.g., "*A is said of all B*") is structurally different from the predicative relation between a predicate letter and an individual variable that occur in the *definiens* (e.g., "for every individual x, if B is said of all x, then A is said of all x"). If one reads carefully *Prior Analytics* A, 1, it is difficult to think that Aristotle is offering mere elucidations of the terms discussed (as Malink [13] claims) and not definitions (for this claim, cf. e.g., [3]). Accordingly, either Aristotle would have proposed a faulty definition of the *dictum de omni* by his own standards[4] – if we were to attribute him the heterodox interpretation – or the orthodox interpretation should be preferred.

But the heterodox reading has its advantages. It nicely fits with Aristotle's theory of predication, and as Malink [10] and [13] showed,[5] the oddities of modal syllogistic – such as the claim that *Barbara NXN* is a valid mood, whereas *Barbara XNN* is not (cf. *Prior Analytics* A, 9) – can be explained away by supposing that the terms occurring in the premises of a syllogism should be the predicables expounded in *Topics* A. In this case, the major necessity premise of *Barbara* NXN would express a predication where the subject is a substance term. If the middle term is a substance term, it will be the predicate of the minor premise, thereby turning the minor into a proposition ultimately expressing a necessity predication, although not displaying a necessity operator in *Barbara*

---

[3]Cf. [13, pp. 35-72], for an evaluation of either alternative. Even though M. Malink is an advocate of the heterodox interpretation, his analysis is balanced and presents many arguments in favor of the orthodox interpretation as well. My summary of the pros and cons of each reading is inevitably selective due to space constraints.

[4]On Aristotle's rejection of circular definitions, cf. *Top.* Z, 4, 142 a34-142 b6.

[5]For an excellent assessment of [13] as a historical reconstruction of Aristotle's syllogistic cf. [15, pp. 63-68]. Cf. [6] and [7] for an overview of contemporary readings of Aristotle's modal syllogistic.

NXN (cf. [10] for details).

If we look at the reception of Aristotle's logic, we can see that Alexander of Aphrodisias was likely a supporter of the heterodox interpretation (cf. [5]), and it seems reasonable to conject that Theophrastus might be responsible for the "heterodox turn" in the Aristotelian tradition after the death of the Stagirite (cf. [4] and [9]), since the philosopher of Eresus argued that each proposition (displaying the structure "P is said of all S") can be translated into its corresponding prosleptic counterpart (e.g., "P is said of all Z of which S is said"). This claim goes hand in hand with the idea that the only types of term that can occur in a proposition are the predicables (i.e., universal terms) and not terms for individuals.[6]

However, it remains controversial to establish whether Aristotle subscribed to the heterodox or to the orthodox version in his writings, especially because it is difficult to suppose that he did not mean to introduce a definition, when he started the sentence by writing λέγομεν δὲ τὸ κατὰ παντός, etc.[7]

---

[6]On Theophrastus logic see [2], [16], [17], and [8].

[7]Malink's reply to this objection is quite elaborate: "Barnes's second objection is that the heterodox *dictum de omni* is of no use as an explanation of the meaning of a_X-propositions, on the grounds that the explanans itself contains a_X-propositions [10, 412]. Thus, a_X predication is explained or defined in terms of a_X-predication, so that the explanation becomes circular. Consequently, the heterodox *dictum de omni* is less informative about the nature of a_X-predication than the orthodox one. The latter provides an explicit definition of a_X-predication in terms of a more primitive notion, namely, that of an individual's falling under a term. This definition allows us to determine the truth of a_X-propositions when we know the extension of the terms involved. The heterodox *dictum de omni* gives no such explicit definition of a_X-predication. Rather, a_X-predication is treated as a primitive, undefined relation. The question of how to determine the truth of a_X-propositions is not answered. Thus, unlike the orthodox *dictum de omni*, the heterodox one does not determine the intuitive meaning of a_X-propositions; it does not define what a_X-predication is.

REPLY TO THE SECOND OBJECTION. Did Aristotle mean the *dictum de omni* as an explicit definition of a_X-predication in terms of a more primitive notion? Morison suggests that the answer is no:

> There is nothing in the language at *APr.* 24b25–30 to suggest that we must construe the *dictum* as a definition. But if it isn't a definition, what is it? The obvious thought is that it is a characterisation of the relations of 'being predicated of every' and 'being predicated of no' in which we are told precisely those facts about the relations which are needed to explain the perfect syllogisms and the conversion rules. [18, 214]

On this view, Aristotle's *dictum de omni et de nullo* is not intended as a definition of what a_X-predication is. Instead, it specifies logical properties of a_X- and e_X-predication that account for the validity of his perfect moods and conversion rules. As such, the heterodox *dictum de omni et de nullo* is informative. It states that, for any A and B, A is a_X-predicated of B if and only if A is a_X-predicated of everything of which B is a_X-predicated. Given classical

2 A controversial text

In my opinion, there is a text that can shed a light on this issue. It is taken from the second book of the *Prior Analytics*, a section largely understudied of Aristotle's treatise on syllogisms:[8]

> When A belongs to the whole of B and to C and is affirmed of nothing else, and B also belongs to every C, it is necessary that A and B should be convertible; for since A is said of B and C only, and B is affirmed both of itself and of C, it is clear that B will be said of everything of which A is said, except A itself. Again when A and B belong to the whole of C, and C is convertible with B, it is necessary that A should belong to every B; for since A belongs to every C, and C to B by conversion, A will belong to every B (Aristotle, *Prior Analytics*, B, 22, 68a16-68a25, Oxford translation revised by J. Barnes) [1]

> Ὅταν δὲ τὸ Α ὅλῳ τῷ Β καὶ τῷ Γ ὑπάρχῃ καὶ μηδενὸς ἄλλου κατηγορῆται, ὑπάρχῃ δὲ καὶ τὸ Β παντὶ τῷ Γ, ἀνάγκη τὸ Α καὶ Β ἀντιστρέφειν· ἐπεὶ γὰρ κατὰ μόνων τῶν Β Γ λέγεται τὸ Α, κατηγορεῖται δὲ τὸ Β καὶ αὐτὸ αὑτοῦ καὶ τοῦ Γ, φανερὸν ὅτι καθ' ὧν τὸ Α, καὶ τὸ Β λεχθήσεται πάντων πλὴν αὐτοῦ τοῦ Α. πάλιν ὅταν τὸ Α καὶ τὸ Β ὅλῳ τῷ Γ ὑπάρχῃ, ἀντιστρέφῃ δὲ τὸ Γ τῷ Β, ἀνάγκη τὸ Α παντὶ τῷ Β ὑπάρχειν· ἐπεὶ γὰρ παντὶ τῷ Γ τὸ Α, τὸ δὲ Γ τῷ Β διὰ τὸ ἀντιστρέφειν, καὶ τὸ Α παντὶ τῷ Β.

---

propositional and quantifier logic, this implies that the relation of a_X-predication is both reflexive and transitive" [13, pp. 65-66]. Contrary to Morison, it seems to me that the words chosen by Aristotle (λέγομεν δὲ etc.) strongly suggest that he was thinking of a definition of what it is to be "said of all".

[8]On this passage, cf. the commentary by N. Strobach (who summarizes Malink's take on the passage): "Der Absatz 68a16–25 enthält zwei Behauptungen mit Beweisen: 68a16–21 und 68a21–25. Die fünf Zeilen des ersten Unterabschnitts, 68a16–21, sind eine der rätselhaftesten Stellen im ganzen Buch II. Sie werfen die Frage auf, was überhaupt mit Umkehrung (ἀντιστρέφειν) genau gemeint ist. Bisher ließ sich davon ausgehen, dass damit sowohl in II 5–7 als auch in II 16 und II 22 das Übergehen von AaB zu BaA gemeint war. Auch der zweite Teil des Abschnitts, 68a21–25, *allein* ließe daran nicht zweifeln. Behauptet wird dort: Aus AaC, BaC und dessen Umkehrung CaB folgt AaB. Die Begründung ist ohne jede weitere Vorsichtsklausel: Aus AaC und CaB folgt AaB. Man wird zunächst darin ohne weiteres einen Barbara-1 sehen.

Problematisch dagegen ist der erste Teil des Abschnitts, 68a16–21. Wir wissen durch ein anonymes Scholium, dass bereits Alexander von Aphrodisias die in 68a16–21 aufgestellte Behauptung für falsch hielt und lieber einen anderen Text gelesen hätte (Brandis (1836),

In this passage, Aristotle offers a new type of conversion[9]. The question we may ask ourselves is why he adds the clause πλὴν αὐτοῦ τοῦ A. Aristotle's point seems to be the following.

1. A is said of all B
2. A is said of all C

194a40–b2). Pseudo-Philoponos gibt sich alle Mühe, erklärt aber die Behauptung zumindest für „höchst erstaunlich" (θαυμάσιον πάνυ, CAG XIII 2, 470, Z. 6, zu 68a16). Smith findet 68a21 „puzzling" (218), Barnes ((2007), 494) die Stelle unvereinbar mit dem, was Aristoteles sonst sagt.

Betrachten wir zunächst die Situation, die in 68a16–18 beschrieben wird, und das Argument, das Aristoteles in 68a19–21 bietet und klammern wir dabei das in a18 behauptete Ergebnis aus! Man hat AaB, AaC und BaC.

Man hat ferner die Zusatzinformation, dass A von nichts anderem als von B und C ausgesagt wird. Damit wird nicht bestritten, dass A von sich selbst ausgesagt wird. Denn in 68a20 wird als ganz selbstverständlich angenommen, dass B von B ausgesagt wird. A wird also von nichts außer A, B und C ausgesagt. Wir haben BaB (zur Bedeutung dieses Ergebnisses vgl. § 6.3, 8.4). Und es spricht nichts dagegen, dass wir auch AaB und CaC haben. Mit BaB, BaC und der Zusatzinformation ist sofort gezeigt: B kommt allem zu, dem A zukommt – außer A selbst. Es ist nützlich, dafür eine Notation zu haben. Es sei deshalb vereinbart:

$XaY^{/Y} =^{def}$. X kommt allem zu, dem Y zukommt, außer Y selbst.

Das „außer" ist dabei ähnlich zu verstehen wie das inklusive „oder": X mag obendrein noch Y zukommen, das ist nicht ausgeschlossen. Aber es ist nicht gesagt.

Dass B A selbst zukommt, wurde an der vorliegenden Stelle nicht gezeigt. Verblüffend ist, dass Aristoteles in 68a18 als Ergebnis genau dieser Überlegungen behauptet, A und B seien umkehrbar. Warum aber sollte das Vorgebrachte dafür bereits hinreichend sein, ohne dass BaA vorliegt? Dies ist allenfalls denkbar, wenn an dieser Stelle die folgende Definition von „umkehrbar" einschlägig ist:

A und B sind genau dann umkehrbar, wenn AaB & $BaA^{/A}$.

Mit Malink (2009) liegt inzwischen eine eingehende Interpretation von 68b16–21 vor, die genau das vertritt. In der von Aristoteles in 68a16–21 beschriebenen Situation fallen alle Individuen, die unter A fallen, auch unter B (diejenigen, die unter C fallen, fallen laut Voraussetzung ja auch unter B). Man braucht, um das zu modellieren, eine Semantik, in der man nicht aufgrund der Extensionsgleichheit von A und B gezwungen ist, mit der Wahrheit von AaB auch die Wahrheit von BaA zu behaupten.

Liest man die a-Prädikation im Sinne des von Malink vorgeschlagenen heterodoxen dictum de omni und modelliert sie mit einer Quasiordnung für a als Grundrelation (§ 8.6), so erreicht man dieses Ergebnis. Die a-Relation ist nicht-extensional, da genau im in 68a16–21 beschriebenen Fall trotz Extensionsgleichheit von A und B mit AaB nicht zugleich BaA wahr ist (Malink (2009), 120); vielmehr ist BaA falsch und lediglich $BaA^{/A}$ wahr." [14, pp. 486-487].

[9]On the interpretation of this controversial passage see [21], that offers a useful summary of the debate. Zanichelli claims that the extensional interpretation should be preferred because it is simpler.

3. B is said of all C

4. A is said of nothing else but B and C

5. Therefore, B is said of all A.

How is it possible to infer that if A is said all B, then B is said of all A, based on the premises laid down by Aristotle? Malink [12] maintains that this passage supports a non-extensional interpretation that would be in line with his overall reading of Aristotle's syllogistic.[10] Malink's claim had been anticipated by Ross [20], *ad loc.*, who maintains that A is a genus, B its only species, and C the only subspecies of B.[11]

A non-extensional reading is tantamount to state that

(i) A is said of all B

can be read, by applying the heterodox *dictum de omni*, as follows:

(i*) A is said all of the terms of which B is said.

Since B is said of itself and of C, A is said of B and of C (and of nothing else).

The converse of (i) – in virtue of this peculiar *conversio simpliciter* - would be

(ii) B is said of all A

which in virtue of the heterodox *dictum* becomes

(ii*) B is said of all of the terms of which A is said.

---

[10]In Malink's reconstruction, "A is a-predicated of B if and only if the a-proposition 'A belongs to all B' is true. The notion of conversion in question is: A converts with B if and only if A is a-predicated of everything of which B is a-predicated including B itself, while B is a-predicated of everything of which A is a-predicated except of A itself. We may call this asymmetric conversion" [12, p. 105].

[11]Cf. [20, p. 480]: "The situation contemplated here is that in which B is the only existing species of a genus A which is notionally wider than B, and C is similarly the only subspecies of the species B. Then, though A is predicable of C as well as of B, it is not wider than but coextensive with B, and B will be predicable of everything of which A is predicable, except A itself (^{a}20-1). It is not predicable of A, because a species is not predicable of its genus (Cat. 2^b21). This is not because a genus is wider than any of its species; for in the present case it is not wider".

Since, by hypothesis, A is said *only* of B and of C, then both A and B are said *only* of B and of C, hence they are convertible.

The clause πλὴν αὐτοῦ τοῦ A would thus be a specification[12] — which is very common in logic[13] — in order to avoid the objection according to which the *intension* of A and of B could be different, under the hypothesis that A is predicated of itself. As B is said of B and of C, it could be said that its notion includes the following predicates: p_1 "said of B", p_2 "said of C". Similarly, also A includes *only* predicates p_1 and p_2, as long as we do not mention that A is predicated of itself. Accordingly, if the *intention*, i.e., the collection of all

---

[12] I owe this idea to Paolo Maffezioli. Even though it is now difficult for me to read the text differently, I should mention that commentators tried to read into this clause more than a mere iteration of what Aristotle mentioned a few lines earlier, i.e., the stipulation that A is said only of B and of C, and *of nothing else* (i.e., including of A itself). Apart from the reconstruction by M. Malink mentioned above, pseudo-Philoponus offers a non-conventional explanation that seems to be the following.
Premise 1. A is said of [all] B
Premise 2. A is said of [all] C
Premise 3 [implicit in pseudo-Philoponus' reconstruction]. A is said only of B and of C.
Premise 4. For every X, X is said of [all] X.
Premise 5. B is said of [all] C. Conclusion. It is possible to infer that B is said of all C thanks to a syllogism where the premises are "B is said of [all] B" (instance of 4) and "B is said of [all] C" (premise 5). In this syllogism, the term "A" does not occur, and the inference in drawn without A (παρεκτὸς τοῦ A τοῦ $Γ$ κατηγορεῖται τὸ B). While logically correct, the commentary of pseudo-Philoponus does not seem to address the law of conversion that Aristotle is seemingly introducing in *Prior Analytics* B, 22, 68a16-68a25. While trivializing Aristotle's claim, the commentary implicitly attests to the controversial nature of the statements made by the Stagirite: Ἕτερον θεώρημα παραδίδωσι θαυμάσιον πάνυ. ἔστι δὲ τοιοῦτον· ἐὰν τὸ A κατηγορῆται καὶ τοῦ B καὶ τοῦ $Γ$, κατηγορῆται δὲ καὶ τὸ B τοῦ $Γ$, εἰδέναι δεῖ ὅτι καὶ τὸ B τοῦ A κατηγορεῖται καὶ ἀντιστρέφει. ἐπειδὴ γὰρ τὸ A κατηγορεῖται τοῦ $BΓ$, <τὸ δὲ B κατηγορεῖται καὶ αὐτὸ ἑαυτοῦ καὶ τοῦ $Γ$> (ἕκαστον γὰρ τῶν ὄντων αὐτὸ ἑαυτοῦ κατηγορεῖται· ὁ γὰρ ἄνθρωπος ἄνθρωπός ἐστιν, καὶ ὁ ἵππος ἵππος ἐστίν), εἰ οὖν τὸ B ἑαυτοῦ καὶ τοῦ $Γ$ κατηγορεῖται, τὸ B ὡς ἑαυτοῦ κατηγορούμενον παντὶ τῷ $Γ$ ἔσται παρεκτὸς τοῦ A. οἷον ἔστω τὸ B ἄνθρωπος, τὸ δὲ $Γ$ ζῷον ἔστω· ὁ ἄνθρωπος ἄνθρωπός ἐστιν, ὁ ἄνθρωπος ζῷόν ἐστιν· οὐκοῦν ὁ ἄνθρωπος διὰ τοῦ 'ἄνθρωπός ἐστι' ζῷόν ἐστιν. ὥστε παρεκτὸς τοῦ A τοῦ $Γ$ κατηγορεῖται τὸ B. καὶ παρεκτὸς πάλιν τοῦ $Γ$ τὸ A κατηγορεῖται τοῦ B διὰ τοῦ κατηγορεῖσθαι αὐτὸ ἑαυτοῦ οὕτως· ἔστω τὸ A γελαστικὸν καὶ τὸ B ἄνθρωπος καὶ ἑαυτοῦ κατηγορείσθω οὕτως· ὁ ἄνθρωπος ἄνθρωπός ἐστιν· ὁ ἄνθρωπος ἄνθρωπος ὢν γελαστικός ἐστιν· ὁ ἄνθρωπος οὖν διὰ μέσου τοῦ 'ἄνθρωπός ἐστιν' γελαστικός ἐστιν. οὕτω καὶ διὰ τὸ αὐτὸ ἑαυτοῦ κατηγορεῖσθαι γίνεται ἀντιστροφὴ ἄνευ τοῦ ἄλλου ὅρου [19, p. 470, ll. 7-21].

[13] As is well known, Aristotle's style is concise, at times to the extreme, but when he writes on logical topics, he is remarkably careful in laying down explicitly all the premises from which he later draws conclusion. As an example of a trivial logical truth (the duality of modal operators), which he takes the time to spell out, we could point to *Posterior Analytics* A 4, 73a21-23: Ἐπεὶ δ'ἀδύνατον ἄλλως ἔχειν οὗ ἔστιν ἐπιστήμη ἁπλῶς, ἀναγκαῖον ἂν εἴη τὸ ἐπιστητὸν τὸ κατὰ τὴν ἀποδεικτικὴν ἐπιστήμην· ἀποδεικτικὴ δ' ἐστὶν ἣν ἔχομεν τῷ ἔχειν ἀπόδειξιν.

determinations that we consider, is $\{p_1, p_2\}$ for both A and B, then A and B are convertible. The proof of the conversion could be the following:

Premise 1. A is$_{by\ deinition}$ $\{p1, p2\}$ [stipulation]
Premise 2. B is$_{by\ deinition}$ $\{p1, p2\}$ [stipulation]
Premise 3. For every X, for every Y, if X is$_{by\ deinition}$ Y, then Y is X.
Premise 4. $\{p1, p2\}$ is A [1, 3]
Premise 5. $\{p1, p2\}$ is B [2, 3]
Premise 6. A is B
Conclusion: B is A [6, 2, 4, Barbara]

However, despite this possible intensional reading of the conversion, it could be possible to read it also extensionally, i.e., by applying the orthodox *dictum de omni*. According to this reading, A and B are convertible if they have the same extension. Since A is said only of all of B and of all of C and of nothing else, and C is included in B, we can assume that there are no individuals that are B but not C, if we also assume that the symmetric difference of the sets B and C is the empty set – a thesis which seems to capture what Aristotle appears to be saying in the passage under discussion.

In other words, with the extensional reading, A and B are convertible because they share the same extension, i.e., "C" (if "C" is an individual) or all the individual x that are included in C (if "C" is interpreted as a class). According to this reading, however, the specification πλὴν αὐτοῦ τοῦ A may appear pleonastic, because if A and B have the same extension (namely C, regardless of the interpretation of C as a class or as a single individual), they are identical and thus convertible (by *conversio simpliciter*). In other words, the specification that A is not said of itself does not appear to be well placed, according to this reading, and Malink [12] was certainly right in noticing it.

3 A possible ambivalent interpretation of the *dictum de omni*

The fact that in one reading the specification πλὴν αὐτοῦ τοῦ A is pleonastic, and in another is not, shows that Aristotle might have felt the need to add it because he wanted to prove the notion of conversion expounded in *Prior Analytics* B 22 not only based on an extensional reading (which would have the clause as pleonastic) but *also* based on the heterodox version of the *dictum*.

In conclusion, I strongly suspect that Aristotle deliberately *chose* to leave ambivalent the reading of the *dictum* in order to accommodate either a set-theoretic interpretation (the so-called "orthodox *dictum de omni et de nullo*") which appears to be the most natural reading of *Prior Analytics* A, 1, where the Stagirite expands *definitions* of the *dictum* and of "being in a whole", or the heterodox intensional reading, that justifies, among other things, the insertion of the clause πλὴν αὐτοῦ τοῦ A in the notion of conversion expounded in *Prior Analytics* B 22. Against this background, it is possible to imagine that Theophrastus chose one of the possible readings, to accommodate his doctrine of prosleptic propositions, thereby influencing later logicians such as Boethus of Sidon and Alexander of Aphrodisias.

This deliberate ambiguity appears to be also the best interpretation of the passage where the *dictum de omni et de nullo* is first introduced. If Aristotle intends to maintain that

Definition 1. *"A is said of all B" iff "A is said of all individuals of which B is said or of all classes of which B is said"*

the notion of "being said of" is not the same in the *definiendum* ("*A* is said of all *B*") and in the first occurrence of the *definiens* ("*A* is said of all individuals of which *B* is said or of all classes of which B is said"). Although this interpretation is arguably not the most natural for today's readers, it has the advantage of presenting *Prior Analytics*, A, 1, 24b27-24b30 as a passage that introduces a definition (as one would expect from the context) and naturally fits with the dialectical nature of Aristotle's method. In fact, Aristotle maintains that predications from genera to species and from species to individuals are transitive (cf. e.g., *Cat.* III, 1b10-15: Ὅταν ἕτερον καθ' ἑτέρου κατηγορῆται ὡς καθ'ὑποκειμένου, ὅσα κατὰ τοῦ κατηγορουμένου λέγεται, πάντα καὶ κατὰ τοῦ ὑποκειμένου ῥηθήσεται· οἷον ἄνθρωπος κατὰ τοῦ τινὸς ἀνθρώπου κατηγορεῖται, τὸ δὲ ζῷον κατὰ τοῦ ἀνθρώπου· οὐκοῦν καὶ κατὰ τοῦ τινὸς ἀνθρώπου τὸ ζῷον κατηγορηθήσεται· ὁ γὰρ τὶς ἄνθρωπος καὶ ἄνθρωπός ἐστι καὶ ζῷον). Additionally, it can be thought that in the context of certain sciences, it could be useful to have chains of predications that descend until the individuals (e.g., in the case of zoology, biology, etc.), whereas in other domains (such as in the case of arithmetic, geometry, etc.), the scientist tends to focus on predications among genera and species. Accordingly, the dialectical games played in each domain are different, and so should be the possible applications of the same *dictum*. In conclusion, the dialectical nature of Aristotle's thinking can explain the ambivalence embedded in his definition of the *dictum de omni et de nullo*.

References

[1] BARNES, J. *Truth, etc.* Oxford, Oxford University Press, 2007.

[2] BOCHEŃSKI, J. *La logique de Théophraste.* Fribourg (CH), Librairie de l'université, 1947.

[3] CRIVELLI, P. *Logic,*. In *C. Schields (ed.), The Oxford Handbook of Aristotle.* Oxford, Oxford University Press, 2012.

[4] GILI, L. Boeto di Sidone e Alessandro di Afrodisia intorno alla sillogistica aristotelica. *Rheinisches Museum für Philologie, 154, pp. 375-397* (2011).

[5] GILI, L. Alexander of Aphrodisias and the Heterodox *dictum de omni et de nullo*. *History and Philosophy of Logic, 36, pp. 114-128* (2015).

[6] GILI, L. Interpreting Aristotle's Modal Syllogistic. *Documenti e studi sulla tradizione filosofica medievale, 26, pp. 1-12* (2015).

[7] GILI, L. La sillogistica del necessario in alcune interpretazioni novecentesche. *Rivista di filosofia neoscolastica, pp. 445-463* (2016).

[8] GILI, L. I. M. Bocheński and Theophrastus' Modal Logic. *Edukacja Filozoficzna, 70, pp. 19-34* (2020).

[9] GILLI, L. ἕτερόν τι τῶν κειμένων. *The Reception of Aristotle's Logic in Late Antiquity.* Lanciano, Carabba, 2024.

[10] MALINK, M. A Reconstruction of Aristotle's Modal Syllogistic. *History and Philosophy of Logic, 27, pp. 95-141* (2006).

[11] MALINK, M. TΩI vs TΩN in Prior Analytics 1.1-22. *Classical Quarterly, 58, pp. 519-536* (2008).

[12] MALINK, M. A Non-Extensional Notion of Conversion in the organon. *Oxford Studies in Ancient Philosophy, 37 pp. 105-141* (2009).

[13] MALINK, M. *Aristotle's Modal Syllogistic.* Cambridge (Mass.), Harvard University Press, 2013.

[14] MALINK, M., AND STROBACH, N. *Aristoteles, Analytica Priora. Buch II, übersetzt von M. Malink und N. Strobach, erläutert von N. Strobach.* Berlin, de Gruyter, 2015.

[15] MCCONAUGHEY, Z. *Aristotle. Science and the Dialectician's Activity. A Dialogical Approach to Aristotle's Logic.* PhD thesis, Université de Lille/Université du Québec à Montréal, 2021.

[16] MIGNUCCI, M. Per una nuova interpretazione della logica modale di Teofrasto. *Vichiana, Vol. 2, pp. 3-53* (1965).

[17] MIGNUCCI, M. Theophrastus' Logic. In *Theophrastus: Reappraising the Sources*, eds. J. van Ophuijsen, M. van Raalte, pp. 39-65. Leiden, Brill, 1998.

[18] MORISON, B. Aristotle, etc. *Phronesis, 53, pp. 209-222* (2008).

[19] PHILOPONI, I. *In Aristotelis Anaytica Priora commentaria,* ed. M. Wallies. Berlin, Reimer, 1905.

[20] ROSS, W. D. *Aristotle, Prior and Posterior Analytics, a revised text with intro-*

duction and commentary by W. D. Ross. Oxford, Clarendon Press, 1964.

[21] ZANICHELLI, R. Notes on Prior Analytics ii 22.68a 16-21. *Elenchos, 44, pp. 81–90* (2023).

Hylomorphism and the Platonization of Logic

Karel Šebela
Department of Philosophy, Faculty of Arts, Palacky University Olomouc
`karel.sebela@pol.cz`

1 Introduction

Since its founding by Aristotle, logic has shown a striking focus on the form of argumentation. This distinction between form and content still plays a key role in logic, not coincidentally called hylomorphism in logic [14]. In modern logic, the emphasis on form has prevailed so much that formal logic has almost become synonymous with logic, and the times when we can find some material logic in logical textbooks seem almost forgotten. In this paper, I will focus on the relationship of this proclaimed formality of logic to the hylomorphic framework. Surprisingly enough, Aristotle is not the father of the hylomorphic doctrine in logic. The question is then when and why this distinction was introduced into logic and if there are any philosophical implications of such a decision. The main thesis of this article is that the Aristotelian dichotomy of matter and form in logic paradoxically became the path to its platonization. Consequently, it will be argued that the doctrine of hylomorphism requires a specific theory of concepts. The main thesis will be argued as follows: first, it will be shown when logical hylomorphism appears in the history of logic and what interpretation is given to it. We will then turn to Kant, who plays a crucial role in our story, since it is he who explicitly conceives of logic as formal, i.e. focused purely on form. We will take a closer look at Kant's argument for the formality of logic. In analyzing it, it will become clear that a crucial part of Kant's argument is a specific theory of concepts, which will thus be understood as a necessary part of logical hylomorphism.

So, the course of the paper is as follows: section 2 brings concise characteristics of logical hylomorphism, sections 3 – 6 yield evidence for the first part of the main thesis, and sections 7 and 8 shows that the doctrine of hylomorphism requires a specific theory of concepts.

2 What is logical hylomorphism?

In [15, pp. 395–396] we can find concise characteristics of logical hylomorphism. According to Dutilh Novaes, there are eight characteristics of the doctrine, for our present purposes it is sufficient to list the following two:

1. In every argument, there is something that corresponds to its form and something that corresponds to its matter.

2. Given that logic is the systematic study of the validity of arguments, it is essentially concerned with forms of arguments.

In sum, logical hylomorphism uses the distinction between matter and form known from Aristotle's philosophy and applies it to the subject matter of logic, *i.e.* to the arguments. To practice hylomorphism in logic, you have to find something analogical to form and something analogical to matter in arguments and focus on the logical form, which alone is responsible for the validity or invalidity of a given argument. Typically the task is accomplished by disparting the vocabulary into two parts depending on whether the term is significant for the inference. If yes, the term becomes a part of the logical form. The rest becomes a part of logical matter and in a formal language is often almost completely omitted "to forgo expressing anything that is without significance for the inferential sequence", as Frege puts it in his Concept Script [9, p. 6]. Surely the very division can be made in different ways and therefore the distinction can be seen as not sharp, so the demarcation of vocabulary belonging to logical form is possibly the greatest weakness of this idea, as Dutilh Novaes points out [15, pp. 393–394]. Nevertheless, the focus on logical form is nowadays something quite common for the practice of logic so it is worth noticing to find its sources and eventual philosophical assumptions.

I think it will be appropriate to conclude this section with the second characteristic of logical hylomorphism. According to it, logical inquiry focuses primarily on the form of arguments. Already in this one could see the difference between, say, classical hylomorphism and logical hylomorphism. It is true that even in classical hylomorphism the emphasis has traditionally been on form rather than matter. But in logical hylomorphism the preference is extreme - matter is understood here only as something to be abstracted from, then only the actual investigation begins. I mentioned in the introduction that in addition to formal logic, there could be and has been for some time a material logic. Thus, for example, David Ross, in his commentary on Aristotle's *Analytics*, characterizes this possible division as follows:

> Formal logic pertains to the *structure* of deduction and proof, with little-to-no reference to content. Material logic pertains to the metaphysical *background*, scientific *content*, and scientific *conditions* of proof. [19, p. 51]

Ross further show the need for material logic:

> Syllogistic inference involves, no doubt, some scientific knowledge, viz. the knowledge that premisses of a certain form entail a conclusion of a certain form. But while formal logic aims simply at knowing the conditions of such entailment, a logic that aims at being a theory of scientific knowledge must do more than this; for the sciences themselves aim at knowing not only relations between propositions but also relations between things, and if the conclusions of inference are to give us such knowledge as this, they must fulfil further conditions than that of following from certain premisses.[1]

Despite these efforts, however, material logic has remained only a half-forgotten chapter in the history of logic and the name is hardly used any more.

3 Plato and Platonism

In the Introduction, I speak about the platonization of logic, so it will be convenient to focus more on the very term and to show its roots in Plato.

According to me, one of the most beautiful and comprehensive characterizations of logic ever written is a passage from Plato's Timaeus, where it is written:

> ...that God devised and bestowed upon us vision to the end that we might behold the revolutions of Reason in the Heaven and use them for the revolvings of the reasoning that is within us, these being akin to those, the perturbable to the imperturbable; and that, through learning and sharing in calculations which are correct by their nature, by imitation of the absolutely unvarying revolutions of the God we might stabilize the variable revolutions within ourselves. (Timaeus 47b-c). [18]

---

[1]Ibid.

There is much contained in this grand picture of Plato, though Plato is not the founder of logic, and the statement is not, on the face of it, about logic. As Plato writes, the arrangement of circulations in us is primarily something non-subjective, because it is an imitation of God's ways. However, it is important for our purposes mainly that imitation is an imitation of a perfect superlunar world, not by imitating the sublunar world.

Therefore, the ideal for logic is a perfect world, existing outside of the ever-changing events around us. This perfect world is then a guarantee not only of the aforementioned non-subjectivity, i.e. the objectivity of logic but also of the immutability and eternal validity of its laws. For the conception of logic as a rigorous science, this way of the foundation is certainly very tempting. It is therefore not surprising that even later in the history of logic, even in the logic of the 20th century, some form of Platonism conceived in this way is still the preferred variant by many logicians.

As for Platonism, it would be useful now to clarify what is meant by this term in the article. Platonism is the view that there exist such things as abstract objects — where an abstract object is an object that does not exist in space or time and which is therefore entirely non-physical and non-mental.[2] By Platonization I mean the tendency to understand a given object of inquiry as a kind of abstract entity, distinct from empirical objects. We must still distinguish here between Platonism and a mere focus on the immaterial. If the proponent of Aristotelianism focuses on form rather than matter, then this does not, of course, mean that he thereby becomes a Platonist. The difference here is the nature of the immaterial. In Aristotelianism, the relation of this immaterial to empirical objects is guaranteed, because the forms are originally in the empirical objects themselves and are abstracted from them, so that independently of these empirical objects they are only in the minds that abstracted them. In Platonism, on the other hand, they have the status of independent entities, for which only subsequently a relation to some empirical objects is possible but not necessary.

My thesis that hylomorphism in logic leads to its Platonization can now be refined to say that the Platonization of logic leads to logic's dissociation from its relation to empirical objects. In the following sections, we will see how the idea of Platonism in logic sneaks into the Aristotelian concept of logic.

---

[2][4].

4 Aristotle

In the whole corpus of Organon, we do not find any mention of the fact that Aristotle would apply the distinction between matter and form to logic. The only mention where Aristotle uses matter/form distinction in connection with logic is in the second book of Physics.

> The letters are the causes of syllables, the material of artificial products, fire, &c., of bodies, the parts of the whole, and the premisses of the conclusion, in sense of 'that from which'. Of these pairs, the one set are causes in the sense of substratum, e.g. the parts, the other set in the sense of essence-the whole and the combination and the form. (Physics, II.3.195a16-21) [3]

There he equates the relationship of the premises to the conclusion to the relationship of matter to form and makes an analogy with the house - just as the stones are necessary to build a house so the premises are necessary to conclude.

Why did Aristotle not use the matter/form distinction in his logic? Of course, we have to consider carefully possible ahistoricity and the element of contingency, but still, I would like to show that the reasons for this could be philosophical. In what follows I will outline the development of logical hylomorphism in the works of peripatetic and Neoplatonist commentators of Aristotle and try to bring evidence for the claim that the introduction of logical hylomorphism is closely connected with the platonization of logic.

5 Commentators

Using the matter/form distinction in logic appears not before the Peripatetics. Barnes claims that as far as he knows "there is no other evidence for the use of matter and form in logical theory before Alexander" [5, p. 75]. So first is Alexander of Aphrodisias, who wrote in his commentary to *Prior Analytics*:

> The figures are like a sort of common mould. You may pour matter into them and shape the same form for different matters. Just as, in the case of moulds, the matters fitted into them differ not in respect of form or figure but in respect of matter, so too is it with the syllogistic figures. [17, 6.16—21].

The use of mould analogy is quite common in Aristotle when it comes to the distinction between form and matter, so in this respect the connexion with Aristotle is obvious. Interestingly, the first use of hylomorphic vocabulary is in the context of the theory of syllogism and concretely what is at the first time taken as logical form are syllogistic figures. Alexander compares syllogistic figures to the common patterns (typos koinos), whereas the conjunction of premises and modus, also belong to a form. According to Alexander of Aphrodisias, the point of schematic letters in Aristotle's logical treatises is

> ...to indicate to us that the conclusions do not depend on the matter but on the figure, on the conjunction of the premises, and on the modes. For so-and-so is deduced syllogistically not because the matter is of such-and-such a kind but because the combination is so-and-so. The letters, then, show that the conclusion will be such-and-such universally, always, and for every assumption. [17, 53.28-54.2]

This is one of the first uses of hylomorphism in logic. Notice that the usage of matter/form distinction is not exactly the same as in the classical hylomorphic doctrine.

An inference that reads that if A is greater than B, B greater than C, then A is greater than C, it is not according to him syllogism because the conclusion can be deduced from the premises "on account of the particularity of the matter" [14, 262]. Alexander tries to justify by this that intuitively we would take the argument as valid, although it does not meet the requirements of the syllogistic figures. It seems that, according to Alexander, a syllogism is only that argument which has the form (sic!) of one from syllogistic figures. What doesn't belong to figures, conjunctions of premises, or modes, belongs to the matter of inference. In the Neoplatonic commentators, we can find similar expressions – Ammonius expressly asserts that

> ...in every syllogism, there is something analogous to matter and something analogous to form. Analogous to matter are the objects (pragmata) themselves by way of which the syllogism is combined, and analogous to form are the figures (schemata) (AnPr 4.9-11) [5, p. 41].

This differentiation suggests that what logic deals with are forms (of a syllogism), which would mean that logic is not interested in things alone. This consequence is explicit in the play when Ammonius thinks that logic is or is not part of philosophy. According to Ammonius, if we consider arguments together with things (pragmaton), eg. syllogisms themselves together with the things that are their subjects, logic is a part of philosophy. But if we consider empty rules separated from things it is an organon (AnPr 10.38-11.3). We can find in Philoponus a similar consideration, only instead of empty rules he works with universal rules (Enn I.iii.5.10-12, cf. 4.18-20 [5]). According to MacFarlane [14, p. 264], this terminology probably comes from Plotinus. For Plotinus compares Platonic dialectic, which is part of philosophy and deals with real things, with Aristotelian logic that provides "empty theorems and rules". According to Ebbesen [8, p. 134-6], for Porphyrius was an Aristotelian logic not incompatible with Platonic metaphysics, because logic says nothing about ontology! So it seems that although hylomorphism was incorporated into logic already by older peripatetics, it was Neoplatonic commentators, who gave hylomorphism in logic the fashion that is still influential today – logic deals purely with the form of inferences (or judgments), factual contents are not of interest to it and if possible are not its subject at all. It is isolated clean form and that (and let's say only that) is the subject of systematic study. Logic is thus deprived of direct relation with real reasoning and things by which it is argued. The advantage is the independence of such examination of the state of the world and also the necessary validity of its knowledge. We can see Plato's theme about the paths of reason in heaven again. As a counterpoint to these tendencies.

I would like to quote researcher Eleonore Stump, who in her final valuation of Boethius' philosophy writes words that are a clear counterpart to the Platonizing tendencies in logic:

> For Boethius dialectic was largely a corollary of metaphysics. The world has a certain nature, in consequence of which certain things are invariably or at least regularly connected with each other. Because we can know this nature and the variable or regular connections it involves, we can know that certain inferences among propositions preserve truth. [20, p. 2]

I think this interpretation will be more in the Aristotelian line, but the history of logic takes different path.

So, I suppose that although hylomorphism in logic is a conceptual tool taken from Aristotelianism, its use leads rather to the Platonic understanding of logic. Hellenistic debates about the nature of logic cannot, given the historical prelude at the end of antiquity, be understood as that they had a direct influence on later concepts of logic. The idea that logic should be formal, was fully established during scholastic discussions.

6 Abelard

I will not examine here the medieval logic at length, I just focus on one important issue. The key role here is Abelard's work. Abelard distinguishes strictly between two types of inferences. First, we have inferences like If every man is an animal, every man is alive. This inference takes its necessity from the nature of things, Abelard named it the imperfect inference.

The other case are inferences like If every man is an animal and every animal is alive, every man is alive. This inference takes its necessity from the structure of the antecedent itself, Abelard dubbed it perfect inference. According to Abelard, the inference is perfect when

> ...from the structure (complexio) of the antecedent itself, the truth of the consequent is manifest, and the construction (constructio) of the antecedent is so disposed that it contains also the construction of the consequent in itself, just as in syllogisms or in conditionals which have the form of syllogisms. [14, p. 281]

For both perfect and imperfect inferences to be a necessary connection between antecedent and consequent. E.g. the inference if man is a species of stone, then if [something] is a man, it is a stone is a perfect inference because it could not take its necessity from the nature of things.

Perfect inferences hold „in all terms", but this is not good enough for Abelard! After the substitution of terms, we have to „look" and see whether the result of the substitution is true – „the inference of a syllogism is supposed to be so perfect that no relation of things pertains to it."

As Dutilh Novaes noticed [15, p. 403], this very distinction is made with no use of explicit hylomorphic terminology. Even the names, perfect and imperfect, are not hylomorphic. But soon after Abelard, this distinction start to be known as the difference between formal and material consequences, fully developed by John Buridan (see [15, p. 404]), whose account of it "is perhaps the clearest and best" [5, p. 52].

Abelard's distinction between perfect and imperfect inferences, and especially his reasoning for this distinction, became canonical. Although the terminology has changed, in logical tradition the fundamental difference between inferences, the validity of which, in a more modern vocabulary, is given a priori and by the others has been preserved since his time. The former assumed that they were valid based on form, and the view spread that these were the very subject of logic, that logic is formal. Dependence on the world is a sign of non-logicality and materiality. It is not necessary for the logic thus cultivated to be Platonist (it may be a version of nominalism), in any case, contact with the "nature of things" (viz Stump's quotation about Boethius above) is not necessary here. Originally Aristotelian dichotomy of matter and form paradoxically became the path to its platonization.

7 Kant

The last step here is Kant. Kant's position here is exceptional, because Kant coins the very term "formal logic", but more importantly Kant not only uses the hylomorphic distinction but he is the first to have attempted to demarcate logic through its formality [15, p. 405]. In Kant, there is an argument (reconstructed in MacFarlane [14, 121-126]) for the thesis that logic is formal (more precisely and in Kant's terms, general logic is formal, not transcendental logic), i.e. that "the universal and necessary rules of thought, in general, can concern merely its form and not in any way its matter". (JL 12) [10], [11].

So let us take a closer look at this argument; its analysis will allow us to better understand the sources of hylomorphism in logic. The argument is based on the following premises:

(*TS*) Thought is intelligible independently of its relation to sensibility.
(*CJ*) Concepts can be used only in judgment.
(*JO*) Judgment essentially involves the subsumption of an object or objects given in intuition under a concept.
(*OS*) Objects can be given to us only in sensibility. That is, for us (as opposed to God), all singular representations are sensible.

In the following, we focus on premise (*CJ*). What are Kant's reasons for such a claim? First of all, Kant is convinced that to think is to judge. This priority of judging in the process of understanding is the background of (*CJ*). A typical place for this claim is Kant's article *The False Subtlety of the Four Syllogistic Figures Proved*.

> *Firstly*, then I would say: a distinct concept is only possible by means of a judgement, while a complete concept is only possible by means of a syllogism. A distinct concept demands, namely, that I should clearly recognize something as a characteristic mark of a thing; but this is a judgement. In order to have a distinct concept of body, I clearly represent to myself impenetrability as a characteristic mark of it. [12, p. 92].

Kant argues here when we have a concept of a thing then to have a clear representation of the concept's marks we have to recognize those marks as belonging to a thing. This recognition is an act of understanding, namely a judgment. Therefore, these acts of judging precede the constitution of a distinct concept. Thus, to grasp a (distinct) concept automatically means to judge. (CJ) is warranted.

Corey W. Dyck [7, pp. 15–21] argued that Kant takes the thesis about the priority of judging from Christian Wolff's logic. In Wolff, one can find a more detailed exposition of the thesis in his theory of universal concept formation. Firstly, according to Wolff, every judgment involves either attributing or removing some mark from a thing. Secondly, when we intuit an object, say a triangle, we can distinguish e.g. three angles as a mark of it. Thirdly, now we can form a universal concept of the triangle as having three angles. The point is that to perform step two, to discern a mark as a mark of the thing means (according to step one) to judge. As Wolff puts it:

> To the extent, then, that you recognize which [notions] agree with things, which ones cannot always agree, to that extent you judge (cited from [7, p. 16]).

Until now, I've focused on the premises of MacFarlane's reconstruction of Kant's argument. From these premises it now follows (LC) General logic abstracts entirely from the content of the concepts. In general logic, according to Kant, the relation of concepts to empirical objects is thus completely cut off. This, let us say, conceptualist view of concepts is thus a prerequisite for understanding logic as exclusively focused on form.

8 Kant vs. Aristotle

Now it is extremely important that premise (CJ) is incompatible with Aristotle's theory of concepts. Aristotle in *On Interpretation* states that "there are in the mind thoughts which do not involve truth or falsity" (16a9-10) [1]. Now obviously to involve truth or falsity means to judge, consequently, it means that according to Aristotle we can think without judging. The idea that we can think without judging is to be found in other significant places in Aristotle's work, e.g. in his treatise *On Soul*, where Aristotle emphasizes that

> the thinking of undivided objects is among those things about which there is no falsity. Where there is both falsity and truth, there is

already a combination of thoughts as forming a unit (430a 27-29) [2].

The act in which the meaning of words uncombined is held before the mind is in the later scholastic logic called the *apprehension simplex* and is widely taken as the first act of intellect in which we deal with concepts. *E.g.*, in Brandom [6] we can read that

> The pre-Kantian tradition took it for granted that the proper order of semantic explanation begins with a doctrine of concepts or terms, divided into singular and general, whose meaningfulness can be grasped independently of and prior to the meaningfulness of judgments. ... Kant rejects this. One of his cardinal innovations is the claim that the fundamental unit of awareness or cognition, the minimum graspable, is the judgment. [6, p. 79].

Therefore, in Aristotle concepts can be used otherwise than only in judgments, which contradicts (CJ).

The reason for this difference is that the very act of conceiving is in Aristotle's doctrine an act that establishes contact with reality. Kant's conceptualism does not presuppose that. Why? As it is well known, Kant distinguishes very sharply between the cognitive faculties of sensibility and intellect. Sensibility (or capacity of receptivity) produces singular representations, and intellect (or capacity of spontaneity) produces general representations, i.e. concepts. Kant describes a concept as a "function" (*Funktion*), which he characterizes as "the unity of the act through which different representations are ordered under a common one" (in the Critique of Pure Reason, A68/B93 [13]). Now, the important thing is that common features of things cannot be given to us in sensibility. In that sense, generality is always made, by the capacity of spontaneity. As Michael Oberst points out, "for Kant the expression "universal thing" is "entirely condemnable," since, in his view, every existing thing can be only singular" [16, p. 388].

So the reason why Aristotle is not the father of logical hylomorphism lies in his realistic theory of concepts and one of the key premises for the formality of logic lies in the conceptualist theory of concepts. As mentioned in Chapter 3, the Platonization of logic mean logic's dissociation from its relation to empirical objects. We can now see that this separation from empirical objects in Kant is due to his specific theory of concepts, so then it serves as part of his argument for the formality of logic.

9 Conclusion

As we have seen, the very hylomorphic division in logic leads to a platonization of logic. So it seems that it was a good move by Aristotle not to employ it in his logical inquiries. Of course, Aristotle is silent about the reasons for this omission, but it remains a historical fact that shortly after the integration of hylomorphic distinction the platonic interpretation of logic starts up. Perhaps it is the very idea of focusing on pure form that arouses Platonist tendencies. In any case, in Kant we have seen another important assumption that leads to the Platonization of logic. A focus on pure form alone would not lead directly to Platonization, but a focus on pure form along with the assumption that it is separate from reality and not part of it, as in Aristotelian ontology. This assumption is found in Kant and is what justifies the premise (CJ) of his argument about the formality of logic.

References

[1] ARISTOTLE. *On Interpretation.* no pub, no year.

[2] ARISTOTLE. *On Soul.* no pub, no year.

[3] ARISTOTLE. *Physics.* no pub, no year.

[4] BALAGUER, M. Platonism in Metaphysics. In *The Stanford Encyclopedia of Philosophy*, E. N. Zalta and U. Nodelman, Eds., Spring 2025 ed. Metaphysics Research Lab, Stanford University, 2025.

[5] BARNES, J. Logical form and logical matter. In *Alberti, A., ed. (1990). Logica, Mente e Persona.* Florence: Leo S. Olschki, 1990.

[6] BRANDOM, R. *Making It Explicit: Reasoning, Representing, and Discursive Commitment.* Cambridge, Mass.: Harvard University Press, 1994.

[7] DYCK, C. The priority of judging: Kant on wolff's general logic. *Estudos Kantianos 4 (2):99-118* (2016).

[8] EBBSEN, S. *Commentators and Commentaries on Aristotle's Sophistici Elenchi. Volume I (The Greek Tradition).* Leiden: E. J. Brill., 1981.

[9] FREGE, G. Concept script, a formal language of pure thought modelled upon that of arithmetic, by s. bauer-mengelberg. In *J. van Heijenoort (ed.), From Frege to Gödel: A Source Book in Mathematical Logic, 1879–1931.* Cambridge, MA: Harvard University Press, 1967.

[10] KANT, I. Jäsche logic (logic: A manual for lectures). ed. g. b. jäsche (1800). In *Kant Ak: IX.* no publisher, 1800.

[11] KANT, I. *Lectures on Logic.* Translated J. Michael Young. Cambridge: Cambridge University Press, 1992.

[12] KANT, I. *Introduction to Logic.* New York: Barnes and Noble, 2005.

[13] KANT, I. *Critique of Pure Reason*. no pub, no year.

[14] MACFARLANE, J. *What Does It Mean to Say That Logic is Formal?* PhD thesis, University of Pittsburgh, 2000.

[15] NOVAES, C. D. Reassessing logical hylomorphism and the demarcation of logical constants. *Synthese 185 (3):387 – 410* (2012).

[16] OBERST, M. History of philosophy quarterly. *History of Philosophy Quarterly 32 (4):335-352* (2015).

[17] OF APHROSIDIAS, A. *On Aristotle's Prior Analytics I.1-7*. Trans. Jonathan Barnes, Susanne Bobzien. 1991.

[18] PLATO. *Timeus*. npo pub, no year.

[19] ROSS, A. . W. D. *Aristotle's Prior and Posterior Analytics*. New York: Facsimiles-Garl. Edited by W. D. Ross, 190.

[20] STUMP, E. *Dialectic and Its Place in the Development of Medieval Logic*. Ithaca, NY: Cornell University Press, 1989.

The Ontological Argument and the Modal Square of Opposition. A Dialogue between a Theist and an Atheist. Is there a Possible Reconciliation?

JUAN MANUEL CAMPOS BENÍTEZ
Benemérita Universidad Autónoma de Puebla
juan.campos@correo.buap.mx

1 Introduction

This is a dialogue between rational people which happen to share some logical background and a different view on religious matters, especially about God's existence. To make sure the reader can follow them and understand their positions, I present a very short logical introduction of terms and symbols used here.

First, propositional logic is presupposed, the logic which works on the so-called connectives. Two or more sentences may be joined by expressions like "and", "if...then", "entails", "either...or...", "neither... nor...". Negation plays an important role also, "no...", "is not the case that...", and so on. For every connective there are rules, usually a pair of, which enable us to join a pair of sentences or to disjoin them as well. They may be called rules of introduction and rules of separation. We have set the things required in this table 1, where the readings might take several forms, but we use the more usual ones.

There will be only one proposition, G (God does exist) though qualified through several modal properties or modal operators. These can be expressed by a Modal Square of Opposition quite like to the ordinary Aristotelian square with its well-known A, E, I and O vertices. We shall equate them with modal operators and get this Figure 1, a modal square.

We should remember that the typical relationships (contrariety, subcontrariety, subalternation and contradiction) hold also here, that's why quantification, for the Aristotelian square contains quantifiers, and modality may be equated in this sense. Equivalences among quantifiers hold also for modal operators.

Connective	Reading	Symbols	Modal symbols
Conjunction	and, both	∧	Necessity □
Disjunction	either, or	∨	Possibility ◇
Conditional	if...then	⊃	
Entailment	Entails	See modal	Strict implication →
Equivalence	equivalent, if and only if	≡	
Negation	not, it's not the case that	∼	

Table 1: Logical connectives and their symbols

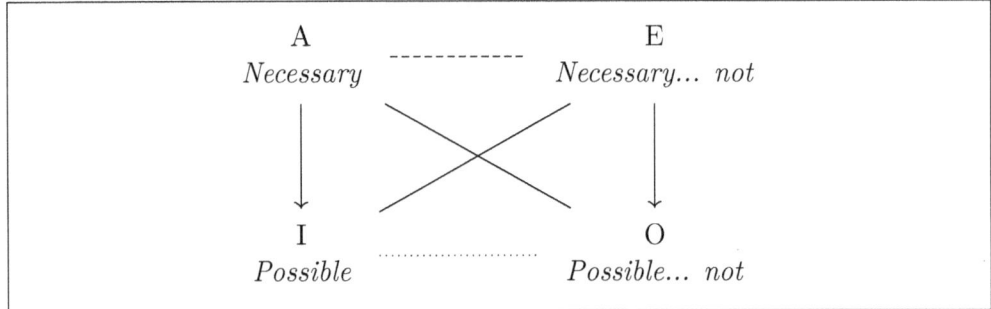

Figure 1: Modal Square of Opposition

$$\Box G \equiv \sim \Diamond \sim G$$
$$\Diamond G \equiv \sim \Box \sim G$$

We may have these readings:

Necessarily G if and only if it is not possible for G to be false.
Possibly G if and only if it is not necessary for G to be false.

According to the Possible Worlds Semantics, we regard a proposition (say G) as a possible proposition when it is true at *some* possible world and take it as necessary when it is true in *every* possible world; impossible when it is true in no possible word, and it is contingent when it is true in *some* possible world and false *in another* possible world. I remark the words which allow us to understand the behavior of modality as a kind of quantification.

Now, we all live in this world, which is called the actual or real world, which 'contains' states of affairs and different events and situations. If we admit that things could have been different from how they really happened, we get an idea of what a possible world could be. In the real world I am giving this talk right now in Sinaia, but something else could have happened, that I didn't arrive on time for example, So, in one possible world I give a talk and into another possible world I do not. We can go further, In one possible world I do exist, and in another one I do not, and so for every man, even for every event and state of affair. Despite this, there are necessary 'things', like the one expressed in this sentence "two plus two makes four" and the claim of the Anselmian argument is that there is a necessary individual which we call God.

Relationships among possible worlds are settled in different modal systems which were first explored by C.I. Lewis; that's why they are called "Lewis' Systems".

The difference among modal systems concerns about their scope regarding necessity. Let me explain this. Suppose a world called w_1 where there is a necessary proposition G. That world can 'reach' another world w_2 which is very close to it. Now, there are worlds w_3 and w_4 closer to w_2 and so on. There's a modal system which allows you to affirm the truth of G in w_2, this is the K system, the weaker of all systems. Another system allows to affirm the *necessity* of G in w_2, and this is the $S4$ system. The next system is $S5$ and allows you to affirm the necessity of G in *every* possible world. It is $S5$ where the necessary existence of God is to be proven.

With this information we can go and listen to the dialogue between an atheist and a theist.

2 Setting the Anselmian Ontological Argument

Saint Anselm's argument admits reformulations, one of which is modal. I will start from this notion close to Hartshorne's argument[1] but I take it from Walter Redmond

Necessarily: if God exists, it is necessary that God does so[2] And from there I will link this notion to the modal square of opposition and equivalence.

Suppose a dialogue between a theist and an atheist, both knowing Sentence logic and the different Modal logic systems.

Theist: I propose

---

[1] See [1, 540].

[2] Necessario: si Deus est, necesse est ut sit. Cfr. [3, 52]. What I will say may be in Redmond's text, perhaps in a somewhat different way, but in this paper I only take that proposition and the theist argument into account, what follows afterwards is all my responsibility.

God exists, entails that it is necessary that God does so.

Atheist: Very well, I will take the contrapositive,

if God doesn't need to exist, God doesn't exist

I can accept that.
Theist: OK, let's put it in symbols

$$G \rightarrow \Box G$$

Where G: God exists, $\Box$: necessarily, it's necessary that...

If G is possible ($\Box G$), then, with the proposed premise, we can conclude $\Box G$ it is necessary for God to exist, in Lewis' $S5$ Modal system

1. $G \rightarrow \Box G$, assumpton

2. $\Diamond G$, assumption
   therefore,

3. $\Box G$, conclusion

Atheist: Right, it follows in $S5$ Lewis' Modal System But I accept this: if God exists, it is necessary that God does so i.e. $(G \supset \Box G)$.
since I accept that if it is not necessary for God to exist, God does not exist

$$\sim \Box G \supset \sim G$$

The consequent is something that interests me, although it is not my priority.
You can notice the difference; you propose a strict implication or entailment while I take the material conditional "if... then" connective.
Your assumption (1) may be defined or rewritten as: $\Box(G \supset \Box G)$ and I just take $(G \supset \Box G)$
And with this premise, plus $\Diamond G$, the conclusion $\Box G$ does not follow.

Theist: True, it doesn't but why you don't take mi premise $\Box(G \supset \Box G)$?

Atheist: Because it seems to me to be an *ad hoc* premise, you need it in order to prove your conclusion, without it you cannot obtain $\Box G$

Theist: I see, you think that if you admit that premise, you lose, because I want to prove $\Box G$, while you want to prove its contradictory, $\sim \Box G$

Atheist: I'm not interested in proving its contradictory, but something stronger, $\sim \Diamond G$, which implies what you say I want to prove...

Theist: Ah! We want to prove propositions with the same modal status, the strongest, i.e. necessary propositions. I want to prove $\Box G$ from the premises $\Box(G \supset \Box G)$ and $\Diamond G$. You admit $(G \supset \Box G)$ since you admit its contrapositive, $(\sim \Box G \supset \sim G)$ but you do not admit necessity for the whole proposition, which is an example of the so-called modality *de dicto*, since you think it is an *ad hoc* strategy so that I can reach the conclusion, the proposition that God exists is necessary...

Atheist: Yes, that is the situation, you need to show why I should admit your premise that seems to me *ad hoc* to your argument...

Theist: If I could show you that my premise is not as arbitrary as you think, would you admit it?

Atheist: If it is not arbitrary, of course I would admit it... but I don't see how...

Theist: Well, let's see. The formula inside the parentheses admits this form $\Box(\sim G \vee \Box G)$. And we can 'put' the square $\Box$ inside the parentheses, qualifying the first disjunct, it could also qualify the second, but it would be redundant to say $(\Box \sim G \vee \Box\Box G)$ since $(\Box\Box G \equiv \Box G)$ is valid in $S4$ Lewis' Modal system and $S4$ is 'contained' in $S5$. So we get

$$(\Box \sim G \vee \Box G)$$

Atheist: Yeah! According to the $S4$ modal theorem $(\Box p \equiv \Box\Box p)$, it's a kind of distribution of the square from the whole to its parts. It is funny.. looks like you changed from one *de dicto* proposition into a *de re* one.
So, what you propose me to admit that your formula leads to this another one, which would give us this formula:

$$\Box(G \supset \Box G) \supset (\Box \sim G \vee \Box G)$$

Theist: It's in fact an equivalence. I think you could admit that God either necessarily exists or necessarily does not. You, as an atheist, shall deny God's existence, and I think you believed that that negation was excluded from my assumption. It was not, given this formula

$$\Box(G \supset \Box G) \equiv (\Box \sim G \vee \Box G)$$

Atheist: ... where one part (from *de dicto* to *de re*) requires the $S5$ modal system and the converse requires the $S4$ modal system. I think we got an agreement, since your formula proposes what we are looking for, since both, you and me, want to prove something that is there. But here we come, we cannot go further. But please continue...

Theist: My argument now goes in this way:

1. $\Box G \vee \Box \sim G$ assumption

2. $\Diamond G$ assumption

3. $\sim \Box \sim G$ 2 equivalence
   Therefore

4. $\Box G$ 1, 3 disjunctive syllogism (a separation rule)

3 The Modal Square of Opposition gets expanded

Atheist: I see that you are proposing a disjunction, and with your extra premise we almost have the modal square of opposition, I mean, three from four nodes. I cannot see how the former formula $\Box(G \supset \Box G)$ fits here. We have the A, E and I nodes in your assumptions, although... the 'universal' nodes joined by a disjunction.

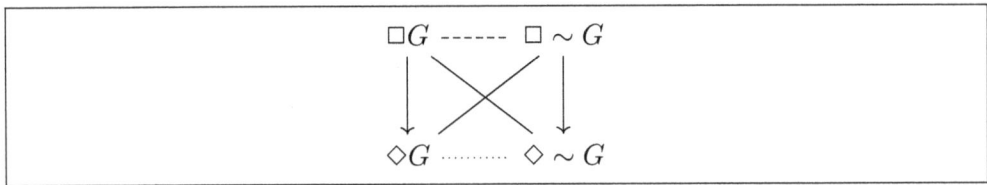

Figure 2: The Modal Square

Perhaps you are taking some advantage of the situation since you have already settled your argument. What about G and $\sim G$? they could be here. As a matter of fact, they are implicitly there, in a kind of Sherwood's hexagon, being subalterns of the universal nodes. Let me call a and e the new members,

I bring your attention to this hexagon because there we have two kinds of contingency, the first is formed by an assertoric sentence in conjunction with a modal sentence ($a \wedge O$ and $e \wedge I$); the second is a conjunction of possibilities ($I \wedge O$):

$$(G \wedge \Diamond \sim G), \text{ or } (\sim G \wedge \Diamond G) \text{ and } (\Diamond G \wedge \Diamond \sim G)$$

I think you will not accept neither of the two forms of contingency, even if you accept the assertoric i.e. non modal part, the a corner

The Ontological Argument and the Modal Square of Opposition

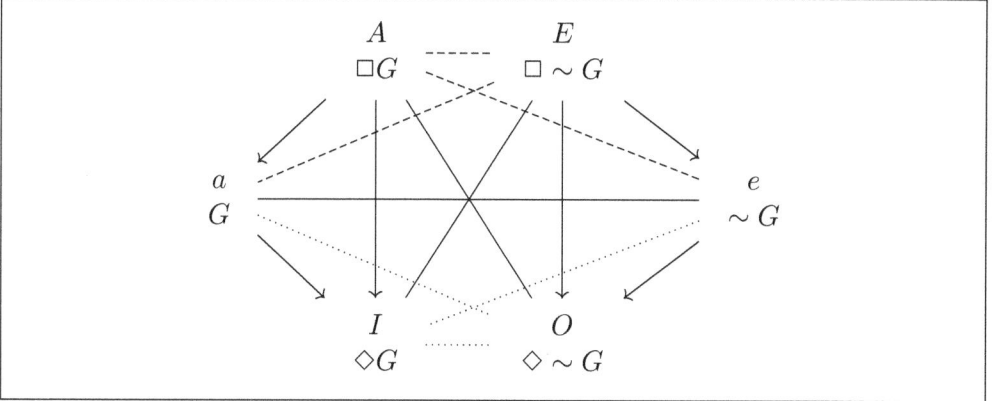

Figure 3: Modal Sherwood's type hexagon

Theist: neither would you accept the contingency even if you accept the assertoric corner, e. Take a look at the following argument:

1. $\Box \sim G \vee \Box G$ assumption
2. $G \wedge \Diamond \sim G$ contingency assumption
3. $\Diamond \sim G$ 2 conjunction separation
4. $\sim \Box G$ 3 equivalence
5. G 2 conjunction separation
6. $\Diamond G$ 5 possibility introduction
7. $\sim \Box \sim G$ 6 equivalence
8. $\Box G$ 1,7 disjunctive syllogism
9. $\Box \sim G$ 1,4 disjunctive syllogism,
   Therefore
10. $\Box G \wedge \Box \sim G$ 8,9 conjunction introduction

We get what we want plus what we do not want. Neither of us can accept any form of contingency here, we agree on that. We also agree on that God is either a necessary or an impossible being, put into another words, the proposition that God exists is either necessary or impossible.

Atheist: Yes, and by accepting your equivalence we share a premise, we both reject contingency, and we are on an equal footing... but I feel that something is still missing...

Theist: We must put all the cards on the table, just as we have made it clear that we must reject divine contingency. I think this Blanche type hexagon puts things in their place:

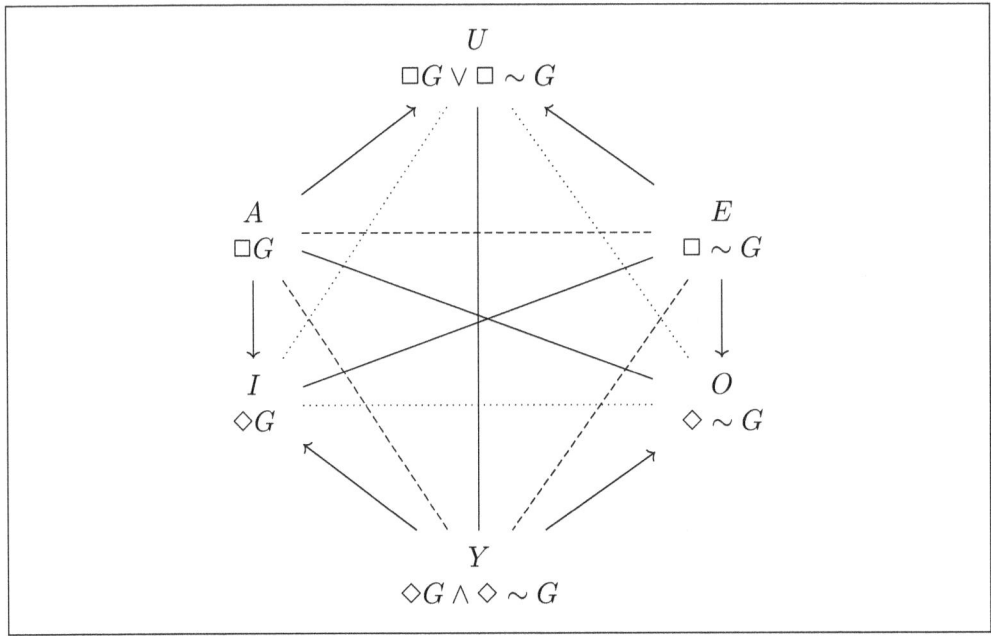

Figure 4: Modal Blanche's type hexagon

If you want to see the whole picture, we should obtain an octagon, to include the assertoric sentences and the compound, i.e. molecular sentences since they include a sentential connective. Notice also that we have arrows going from down to up, something unusual in the traditional Square. We have it twice, from Y to I and O, and from A and E to U. Please notice also that the U's equivalent *de dicto* sentence would seem a little funny here:

4 An Argument from the Atheist and a Common Strategy

Atheist: I wonder if the Y vertex could have a *de dicto* equivalent as it does the U vertex... but that's not a problem here.

The Ontological Argument and the Modal Square of Opposition

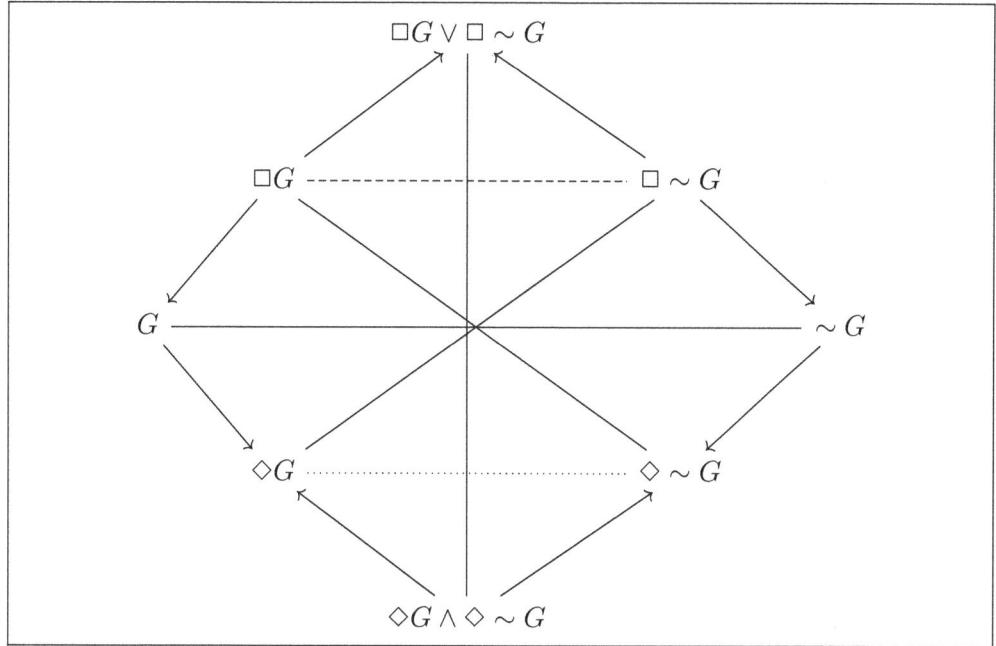

Figure 5: Combining the hexagons into an octagonal figure (Beziau's Octagon)

We have the whole picture now. Notice that both of us follow the same strategy to get our conclusion. You take the U corner plus the I one to get the A sentence.

This being the case, I can offer this argument, which has the same logical form as yours, the only difference being a simple negation at the level of the subcontrary sentences

1. $\Box G \vee \Box \sim G$ assumption
2. $\Diamond \sim G$ assumption
3. $\sim \Box G$ 2 equivalence
   Therefore
4. $\Box \sim G$ 1, 3 disjunctive syllogism

I also take the U corner plus the particular one, but following the negative side, the O to obtain the E sentence. It seems to me we have but a little disagreement, at the level of subcontrary sentences. Little but strong enough to make none of us move into another choice.

We both move in two directions or senses since the premises come from

1) From below, i.e. the subcontrary corner to the universal one
2) From the top, the U corner to obtain the universal one

This diagram shows the common strategy of our arguments, the colored arrows are green to yours and blue to mine. Hick ends his argument at the a corner taking only the left side of the square with no attention to the negative side [?, p. 540]; of course he does not have to do it since the thinks the argument fails. Anyway, his strategy follows this scheme which requires an octagon:

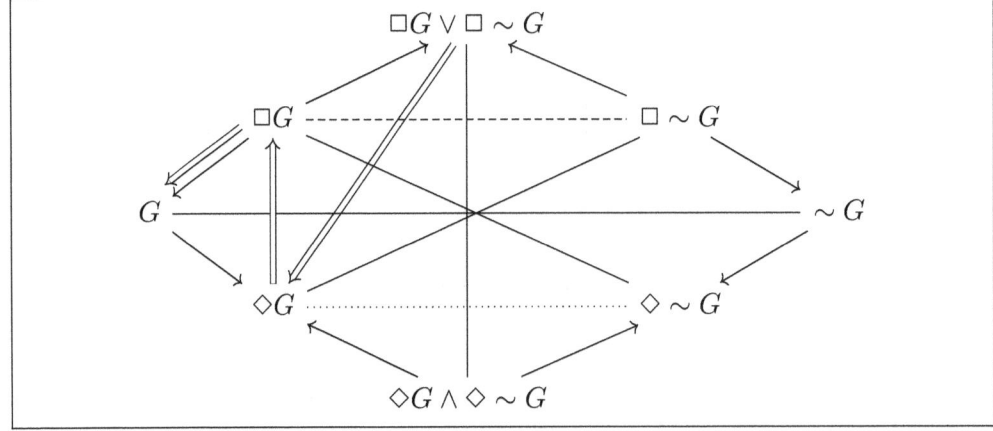

Figure 6: Argument's route

While ours can be set in the following figure I just want to draw attention to the negative side of the square.

It seems to me that from here there is no agreement between you and me, the last battle, should be at the subcontrary level. Subcontraries here neither are nor can be true at the same time, they almost behave as contradictories. Perhaps we are dealing with a different square...

Theist: What do you mean? Different from the Aristotelian square? Well... it is true! Since the subcontraries cannot be joined together... it sounds familiar to me... let me recall.

The Ontological Argument and the Modal Square of Opposition

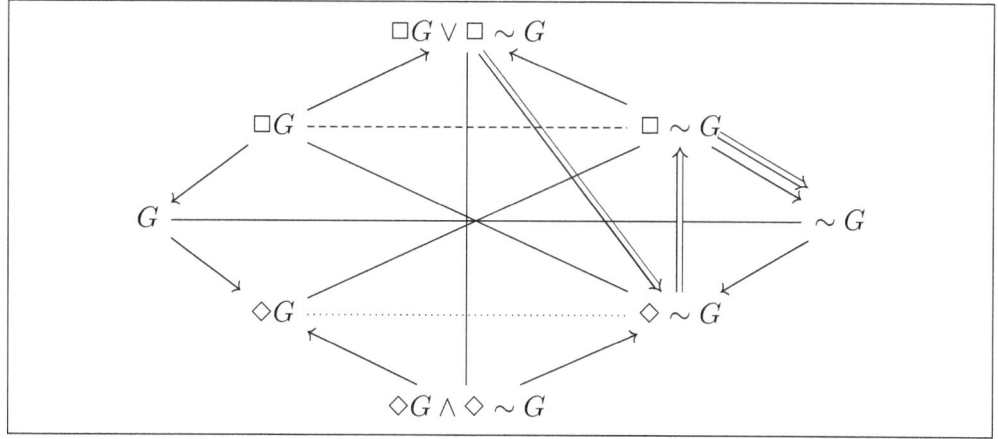

Figure 7: Argument's route 2

5 A Turning on Medieval logicians and Further Problems

Certainly, we have what Medieval logicians called a Square concerning the 'matter' of a proposition, i.e. its modal status. Sherwood takes three modes: necessity or natural matter, impossibility or separate matter, and contingency. Listen to what he says:

> Notice that whenever a particular statement is true in natural matter its subcontrary cannot be true (...). Moreover, in natural matter and in separate matter a particular interchanges with (*convertitur cum*) a universal. Therefore subcontraries in these two matters cannot be true at the same time, and the truth of particular subaltern entails (*infert*) the truth of the subalternant universal. [2, 33 – 34]

Atheist: Alas! That's what we both were doing, going from a particular statement toward a universal one, just as Sherwood puts it, but we need an extra premise though. So, we don't have to abandon the Square in this dialogue. I realize that this fits better our arguments, no wonder since the Ontological Argument comes from a Medieval thinker. Now I have no complains about the logical matters... perhaps but one; actually, it is a doubt about possibility.
Theist: What's it about?
Atheist: I remember my old readings of Medieval philosophy in the 13th Century; they tend to equate possibility with contingency when talking about modalities. If I recall well, the Pseudo-Aquinas says that we should know

that possible and contingent mean the same thing[3]. And Sherwood says that "it is important to realize that the two modes 'possible' and 'contingent' are interchangeable" [2, 47]. This spoils everything for us!

Logicians nowadays take contingency as the conjunction of subcontraries, as we have done. We also have noticed that possibility is our card that will give us the win, but things are not so simple now that we have turned to the Medieval logicians. So, it means we have a great task concerning the notion of possibility and to understand why Medieval men take it the way they do. Probably there will be a sense in which we can both pursue our argumentation. If you agree to this, we can be reconciled for a while as our task and research into the possibility will not be easy. I still believe that the crux of our reasoning is step two, possibility.

If we were to persuade an audience (please allow me to talk in this way), anyone who presents a better defense and justification of the second premise will have won the case... We had had agreements and disagreements, contingency is expelled as a whole, but each one of us takes one single part, you the affirmative side and I the negative one. I do not see any possible reconciliation here. We cannot go further... can we?

Theist: O.K. let's have a truce in order to learn more about possibility, probably metaphysics of modality. When we were done, we will come back to our quarrel.

Even when we are clear about it, and present our defense, I have the feeling that we are forgetting something... something important about our epistemic situation concerning our premises. Let us consider the first premise, are we sure about that?

Atheist: I see what you mean. To be sure, it is not a tautology, so we cannot say we know it is true for it could be false.

And we cannot take the second one as a tautology either... but certainly we believe it is true though its defense is waiting for.

So we have the same doxastic situation

The agent a believes	Premise 1
The agent a believes	Premise 2
The conclusion must be also believed:	
The agent a believes	Conclusion

---

[3]Unde sciendum quod possibile et contingens idem significat, in Corpus Thomisticum, Sancti Thomae de Aquino, De propositionibus modalibus, authenticitate dubium, Avalaible at https://www.corpusthomisticum.org/dpp.html, captured on August 10, 2023.

Theist: that brings up an unexpected consequence. We may both be wrong since belief is consistent with falsity!

Atheist: We should stop our agreements... for what you say is of no help to nobody.

On a second thought... if I accept that both of us could be wrong it only means that we believe but do not know the looked-for proposition. But even if premises being not known, they are **rationally** believed.

Theist: I am not sure we both could be wrong in this issue. Meanwhile I keep my hope in showing you that $\Box G$, even if I'm wrong, I see that you also have problems, similar to mine.

... If we are both wrong perhaps there is room for us in the divine mind.

Atheist: ... But it should be also in this life, please.

THE END?

References

[1] HICK, J. Ontological argument for the existence of god. In *Paul Edwards (Ed.) The Encyclopedia of Philosophy, vol. 5*. New York, Mcmillan Inc., 1967.

[2] KRETZMANN, N. *William of Sherwood's introduction to logic*. U of Minnesota Press, 1966.

[3] REDMOND, W. *Deus et Logica. Logica theologiae philosophicae insita*. México: UPAEP-Porrúa, 2014.

A Logic for Prophetic Conditionals: Perfect Prophecy and Prophetic Intuition

José David García-Cruz
Pontificia Universidad Católica de Chile

1 Introduction

This paper introduces the logic BTEP[1], a multimodal system[2] designed to reason about prophetic information. It features a three-dimensional semantics, which draws from a specific thesis found in Ockham's treatise *De praedestinatione et de praescientia Dei respectu futurorum contingentium* [4]. The logic incorporates the branching conception of time[3], and a conditional approach to understanding prophecies[4]. Additionally, the paper engages with certain elements of Thomas Aquinas' *Summa Theologiae* [11], particularly his distinction between perfect prophecy and prophetic intuition.

The central question outlined in this paper is: What role does the prophet, as an epistemic agent, play in shaping the relationship between divine and human knowledge of the future? This question is significant for at least two reasons. First, it allows us to position prophetic discourse within an argumentative framework, textitasizing the formal properties of human knowledge in relation to information about the future. This, in turn, helps explain why, although prophetic discourse contains necessary elements, it also involves contingent considerations. Second, the question underscores the importance of the prophet as an epistemic agent when reasoning about prophetic discourse. A standard modal logic proves inadequate for capturing the dynamism of agents of this kind.

The structure of the paper is as follows. In the second section, the Ockham and Aquinas' positions are presented. The third section will define the logic

---

[1] *Branching-Time Epistemic Prophecies.*
[2] See [10], [7], and [2].
[3] See [17], [16], [15], [9], and [6].
[4] See [5], [3], [8], and [14].

BTEP, with a particular focus on its three-dimensional semantics. In the final section, several models of prophecy will be analyzed: the first will outline the *prophecy of Nineveh*, while the second will examine the distinction between *prophetic intuition* and *perfect prophecy*.

2 Some Philosophical Elements of Prophecies in Ockham and Aquinas

This section outlines the two philosophical influences that have shaped the development of the logic BTEP. First, we examine Ockham's conception of prophecies as conditionals and present our formulation of his thesis. Second, we describe Aquinas' distinction between prophetic intuition and perfect prophecy.

2.1 Ockham on Prophecies

In this section, we will briefly discuss the conditional analysis of the concept of prophecy, as presented by William of Ockham in his Tractatus de Praedestinatione et de Praescientia Dei respectu futurorum contingentium [4]. Ockham's thesis is framed within the well-known problem of future contingents and their relationship to divine foreknowledge. Ockham argues[5] that two equally plausible theses emerge from this debate:

(1) God knows all future contingents

(2) Human beings can choose between alternative possibilities

The problem arises when both factors are considered together. If God knows all future contingents, then human choice becomes irrelevant, as the future is determined by God's knowledge of which future contingent will be realized. On the other hand, if the human being can choose between alternative possibilities, exercising free will, God's foreknowledge seems to be at odds with these choices. God must know what the human being will choose, but if God knows what the human choice will be, can the human still exercise free will?

Ockham argues that both theses can be reconciled if we consider certain questions related to how we acquire knowledge of the future. He introduces the concept of "communication," specifically, "communication from God to the human being." God communicates the truth about the future, but this does not imply that the statement about the future is necessary. The means by

---
[5]Cf. [13, p. 97].

which God communicates information about the future is through prophecies. Ockham's example is: "Yet forty days, and Nineveh shall be overthrown."[6] For Ockham, the information conveyed through this prophecy about Nineveh is not necessary, but, like all future events, contingent. From this, it does not follow that we cannot access this information in a particular way. Ockham's strategy is to uphold the following thesis:

(OT) All prophecies regarding any future contingents were conditionals.[12, p. 44]

In this sense, prophecies about future contingents take the form of conditionals, where the antecedent remains implicit within the consequent. Ockham argues that we must understand the prophecy of Jonah by presupposing the condition "unless the citizens of Nineveh repent,"[13, p. 97] as the citizens of Nineveh themselves did. This implies that the only way in which God reveals the truth about the future to human beings is conditionally, asserting that a certain situation will occur if certain relevant conditions are met. Ockham maintains that if God were to reveal the truth through unconditional statements, the future would become inevitable, as the revelation would necessarily be true.[7] In this regard, we can say that God reveals information about the future in a "codified" way, using conditionals whose antecedent is implicit in the prophecy but not explicitly stated. The key to Ockham's analysis lies in the distinction between the following two formulations of Nineveh's prophecy:

(CP) Yet forty days, and Nineveh shall be overthrown, unless the citizens of Nineveh repent.

(AP) Yet forty days, and Nineveh shall be overthrown.

At first glance, the only difference between the two formulations is that the first is a conditional, while the second is an atomic proposition. However, this does not indicate which type of conditional is involved, nor does it clarify how to interpret the temporal expressions in each proposition. Since the context of the discussion is temporal, we can assume that the conditional must have a temporal commitment. Furthermore, using the concept of communication, we can infer that the conditional may also be epistemic. Our formulation of Ockham's thesis incorporates these three elements: conditionalization, time, and knowledge.

---

[6][1, Jonah, 3: 4].
[7][13, p. 97].

Therefore, the main idea is that prophecies are epistemic conditionals referring to the future. This conception allows us to reconcile statements (1) and (2) above. God—knowing all contingent futures—reveals information conditionally, which allows human beings to act in such a way that a specific consequent is actualized. In the case of Nineveh, God reveals information conditionally to provoke a change in the behavior of the faithful, thereby preventing the destruction of the city. This does not negate the fact that God knows of alternatives in which the city was destroyed, nor does it conflict with human actions, as these alternatives are caused by human decisions. In the following section, we will formalize these intuitions and revisit this case of prophecy. For now, we will examine Thomas Aquinas's distinction between perfect prophecy and prophetic intuition.

2.2 Aquinas' on Perfect Prophecy and Prophetic Intuition

Within the *Summa Theologiae*, Thomas Aquinas devotes a specific treatise to the question of prophecy, situated in the *Secunda Secundae* [11, II-II, qq. 171-174]. This treatise is part of his broader discussion on the virtues and gifts of the Holy Spirit, specifically addressing how divine knowledge is communicated to human beings. Aquinas examines prophecy in relation to divine omniscience, human cognition, and the metaphysical status of future contingents. His analysis unfolds across several key questions: he first establishes the nature of prophecy (*quid sit prophetia*), then explores its causes, the mode of reception by the prophet, and finally the role of free will and divine illumination in the prophetic act. Central to his account is the idea that prophecy is not merely foresight of the future but a supernatural knowledge granted by divine inspiration, which can manifest in different ways depending on the prophet's disposition and the means through which God imparts knowledge.

Aquinas defines prophecy as a form of knowledge granted by divine revelation, distinct from human reasoning or natural cognition. He states: "Prophecy implies a certain knowledge inspired by God."[8] This knowledge does not arise from rational inference or sensory experience but is directly infused by God. As he further explains: "For man is made perfect in knowing the truth revealed by God through prophecy, not according to the order of nature, but according to the gift of grace."[9] Thus, prophecy belongs to the domain of lumen gratuitum,

---

[8]"Prophetia importat quamdam cognitionem divinitus inspiratam". [11, II-II, q. 171, a. 1].

[9]"Homo enim ad cognoscendum veritatem divinitus revelatam per prophetiam, non perficitur secundum naturae ordinem, sed secundum donum gratiae". [11, II-II, q. 171, a. 1].

a supernatural illumination that enables the prophet to perceive truths beyond the reach of natural intellect.

Aquinas further clarifies that prophecy is principally concerned with future events, yet it can also pertain to present or past realities inaccessible to ordinary human knowledge. He states: "Prophecy pertains principally to the knowledge of future contingents, but secondarily to things present or past, insofar as they exceed human cognition.[10] The prophetic act, therefore, is not limited to foretelling but encompasses any divinely revealed knowledge beyond human grasp. This distinction reinforces the idea that prophecy is not merely predictive but revelatory, participating in the divine intellect's comprehensive knowledge of all temporal realities.

Aquinas establishes a crucial distinction between *perfect prophecy* and *prophetic intuition*, textitasizing the prophet's awareness of divine inspiration. He asserts: "accordingly, when a man knows that he is being moved by the Holy Ghost to think something, or signify something by word or deed, this belongs properly to prophecy; whereas when he is moved, without his knowing it, this is not perfect prophecy, but a prophetic intuition."[11] This distinction hinges on the prophet's cognitive participation: in perfect prophecy, the individual is fully conscious of divine influence, whereas in prophetic intuition, divine motion occurs without the prophet's explicit awareness.

Furthermore, Aquinas highlights the limitations of the human intellect in interpreting divine revelations. Even true prophets do not fully grasp the meaning of their visions or words, as he explains: "Nevertheless it must be observed that since the prophet's mind is a defective instrument, as stated above, even true prophets know not all that the Holy Ghost means by the things they see, or speak, or even do."[12] This indicates that prophetic knowledge, though divinely inspired, remains imperfect and mediated through human faculties. The prophet does not always perceive the full extent of divine revelation, which aligns with Aquinas' broader epistemological framework, where human cognition, even in its highest forms, is finite compared to divine omniscience.

---

[10]"Ad prophetiam autem pertinet praecipue futura contingentia inspicere, secundario autem ea quae sunt in praesenti vel praeterito, inquantum excedunt hominis cognitionem." [11, II-II, q. 171, a. 3].

[11]"Secundum quod, cum homo scit se moveri a Spiritu Sancto ad cogitandum aliquid, aut significandum aliquid verbo vel facto, hoc proprie pertinet ad prophetiam; quando vero movetur, nesciente ipso, hoc non est perfecta prophetia, sed intuitionus propheticus." [11, II-II, q. 173, a. 4]

[12]"Nihilominus, attendendum est quod, quia mens prophetae est instrumentum defectuosum, ut supra dictum est, etiam veri prophetae non sciunt omnia quae Spiritus Sanctus vult per ea quae vident, aut loquuntur, aut etiam faciunt." [11, II-II, q. 173, a. 4]

Aquinas' distinction between perfect prophecy and prophetic intuition aligns with Ockham's view that prophecies about the future are inherently conditional. Since even true prophets do not fully comprehend the divine message, the knowledge imparted through prophecy cannot be absolute but must instead be contingent upon certain conditions. This interpretation resonates with Ockham's idea that prophetic statements, such as Jonah's proclamation regarding Nineveh, implicitly contain an unstated antecedent—unless the people repent. Thus, both thinkers converge in their understanding that divine revelation concerning the future is not an assertion of necessity but rather an invitation to human action. By framing prophecy as conditional, Aquinas and Ockham provide a theological basis for reconciling divine foreknowledge with human freedom, setting the stage for a synthesis of their perspectives. In the following section, we will formalize these insights to further explore their implications.

3 The Logical System BTEP

In this section, we introduce the logical system BTEP, which formalizes the insights derived from Ockham's conditional analysis of prophecy and Aquinas' distinction between prophetic intuition and perfect prophecy. The system is designed to capture the epistemic and temporal aspects of prophecy, ensuring that prophetic statements are treated as conditional assertions about the future. By incorporating modal operators for time and knowledge, BTEP provides a formal framework to reconcile divine foreknowledge with human free will. We begin by outlining the syntax and semantics of the system, followed by an analysis of some models.

Definition 3.1. *The language $\mathcal{L}$ of the system* BTEP *is defined by the following components:*

- *A countable set of propositional variables $Var = \{P, Q, R, \dots\}$.*
- *A set of propositional connectives $\mathbb{C} = \{\neg, \wedge, \vee, \rightarrow, \equiv\}$.*
- *A set of modal operators $O = \{[H], [P], [F], K_a, K_a^\varphi\}$.*

The formulas of $\mathcal{L}$ are generated by the following BNF rule:

$$\varphi ::= P \mid \neg \varphi \mid \varphi \wedge \varphi \mid [H]\varphi \mid [P]\varphi \mid [F]\varphi \mid K_a\varphi \mid K_a^\varphi \varphi.$$

The remaining connectives and modal operators are defined as usual: in the case of connectives, all can be expressed using $\neg$ and $\vee$; in the case of modal operators, all are derivable from ¬ and o, where $o \in O$. The intuitive interpretation of the temporal modalities is as follows. The formula $[F]\varphi$ is read as "φ will always be true in the future," while $[P]\varphi$ means "φ has always been true in the past." Their dual modalities, $\langle F \rangle$ and $\langle P \rangle$, correspond to existential quantification over time: $\langle F \rangle \varphi$ means that "φ will be true at some point in the future," and $\langle P \rangle \varphi$ means that "φ was true at some point in the past."

Historical modalities $[H]$ and $\langle H \rangle$ extend this temporal framework to multiple possible histories. The formula $[H]\varphi$ is interpreted as "φ is true in every possible history," while $\langle H \rangle \varphi$ states that "φ is true in at least one possible history."

Regarding epistemic modalities, the operator K_a represents standard knowledge: $K_a \varphi$ means "agent a knows that φ." The dual operator $\hat{K}_a$ expresses epistemic consistency, meaning that φ is not ruled out by agent a's knowledge. Finally, conditional epistemic modalities provide a finer distinction: $K_a^\psi \varphi$ means that "agent a knows that φ, given that ψ holds," while $\hat{K}_a^\psi \varphi$ expresses that "φ is consistent with agent a's knowledge, given that ψ holds."

To illustrate the expressive power of this language, we now present some examples.

Example 3.1. *The proposition "Alberto knows that at some point in the past, Juan knew the combination of the safe box" can be formalized as:*

$$K_a \langle P \rangle K_j S.$$

Example 3.2. *The proposition "Juan knows the combination of the safe box under the condition that in any possible future history, at some future time, Alberto will know it" can be formalized as:*

$$K_j^{[H]\langle F \rangle K_a S} S.$$

Example 3.3. *The proposition "Augustine knows that time exists if and only if Juan asks him, but he does not know whether time exists if Juan does not ask him" can be formalized as:*

$$K_a^{\neg Q} T \wedge \neg K_a^Q T.$$

Example 3.4. *The proposition "If Juan knows the combination of the safe now, then he will necessarily still know it in the future" can be formalized as:*

$$K_j S \to [F] K_j S.$$

Example 3.5. *The proposition "It is possible that in some alternative history, Alberto never knows the combination of the safe" can be formalized as:*

$$\langle H \rangle \neg K_a S.$$

Example 3.6. *The proposition "Alberto knows that Juan will eventually learn the combination of the safe, under the condition that Juan asks about it" can be formalized as:*

$$K_a^Q \langle F \rangle K_j S.$$

Having established the syntax of our logical system, we now turn to its semantics. The formal language $\mathcal{L}$ we have defined provides a structured way to express statements about time, knowledge, and conditional knowledge. However, to rigorously interpret these expressions, we must specify the underlying models that determine their truth values. In the following, we introduce a class of Kripke-style models that capture the temporal, epistemic, and conditional aspects of our framework. These models will allow us to formally define the meaning of our modal operators and provide the foundation for representing Ockham and Aquinas' ideas.

Definition 3.2. *A model for the language $\mathcal{L}$ is a triple $\mathbb{M} = (W, R, V)$, where:*

- *W is a non-empty set, often referred to as the domain of $\mathbb{M}$, representing the set of possible worlds or states of affairs.*

- *R is a binary relation on W, typically called the accessibility relation, which encodes the way in which worlds or states can relate to one another in the context of the modal operators.*

- *V is a valuation function (or interpretation function), which assigns to each propositional variable $p \in Var$ a subset $V(p) \subseteq W$, representing the set of worlds in which p is true.*

Formally, the model $\mathbb{M}$ satisfies the following conditions:

1. *$W = \{w_i, w_j, w_k, \ldots\} = DOM(\mathbb{M})$.*

2. *$R \subseteq W \times W$.*

3. *$V : Var \to \mathcal{P}(W)$, where $\mathcal{P}(W)$ denotes the power set of W (the set of all subsets of W).*

Definition 3.3. *An epistemic model for the language $\mathcal{L}$ is a tuple $\mathbb{M}_E = (W, A, R, V)$, where:*

- *W is a non-empty set, as in Definition 3.6, representing the domain of the model, i.e., the set of possible worlds or states of affairs.*

- *A is a set of agents, denoted $A = \{a_1, a_2, a_3, \dots\}$, each of which possesses a certain epistemic state or knowledge in the model.*

- *R is a function, called the accessibility relation function, that assigns to each agent $a \in A$ a binary relation $R_a \subseteq W \times W$, representing the accessibility relation for agent a (i.e., which worlds are epistemically accessible to a).*

- *V is a valuation function, as in Definition 3.2, which assigns to each propositional variable $p \in Var$ a subset $V(p) \subseteq W$.*

Formally, the epistemic model $\mathbb{M}_E$ satisfies the following conditions:

1. *$W = \{w_i, w_j, w_k, \dots\} = DOM(\mathbb{M}_E)$.*

2. *$R : A \to \mathcal{P}(W \times W)$.*

3. *$V : Var \to \mathcal{P}(W)$, as in Definition 3.2.*

Note that an epistemic model is essentially a modal model with an additional structure that accounts for the knowledge of multiple agents, represented through distinct accessibility relations. For simplicity, we will use the notation $\mathbb{M}_3 = (W, R, V)$ when the set of agents A is either implicit or clear from the context. The next definition introduces the framework needed to formally characterize temporal models.

Definition 3.4. *A bidirectional model for the language $\mathcal{L}$ is a tuple $\mathbb{M}_T = (W, R, \check{R}, V)$, where:*

- *W is a non-empty set of possible worlds, i.e., the domain of $\mathbb{M}_T$.*

- *$R \subseteq W \times W$ is a binary accessibility relation.*

- *$\check{R}$ is the converse relation of R, formally defined as $\hat{R} = \{(y, x) \mid (x, y) \in R\}$.*

- *$V : Var \to \mathcal{P}(W)$ is a valuation function assigning to each propositional variable $p \in \mathbb{V}$ a set of worlds $V(p) \subseteq W$ where p holds.*

Note that a bidirectional model is essentially a modal model augmented with the converse accessibility relation for R. For temporal models, we will use the notation $\mathbb{M}_T = (T, <, V)$, where $<$ represents a temporal ordering[13]. Similarly, for historical models, we will use $\mathbb{M}_H = (H, \sqsubseteq, V)$, where $\sqsubseteq$ denotes a branching historical structure. Next, we define some characteristic classes of models.

Definition 3.5. *Let* $\mathbb{M} = (W, R, V)$ *be a model for the language* $\mathcal{L}$. *The relation* R *is said to be:*

- **Serial** *if* $\forall w \in W, \exists v \in W$ *such that* Rwv.

- **Reflexive** *if* $\forall w \in W, Rww$.

- **Transitive** *if* $\forall w, v, u \in W$, $(Rwv \land Rvu) \to Rwu$.

- **Symmetric** *if* $\forall w, v \in W$, $Rwv \to Rvw$.

- **Euclidean** *if* $\forall w, v, u \in W$, $(Rwv \land Rwu) \to Rvu$.

- **An equivalence relation** *if it is reflexive, transitive, and symmetric.*

These constraints correspond to the standard conditions used to characterize the most common modal logic systems. The class of serial models is denoted by $\mathcal{KD}$, the class of reflexive models by $\mathcal{T}$, and the class of transitive models by $\mathcal{K}4$. The class of models that are both reflexive and transitive is denoted by $\mathcal{S}4$. Symmetric models are represented by $\mathcal{B}$, while Euclidean models are denoted by $\mathcal{K}45$. The class of serial, transitive, and Euclidean models is denoted by $\mathcal{KD}45$. Finally, models satisfying the properties of an equivalence relation (i.e., reflexivity, transitivity, and symmetry) are denoted by $\mathcal{S}5$.

This classification applies to the three types of models introduced so far: temporal, epistemic, and historical. With these foundations in place, we can now formally define product models[14] for the language $\mathcal{L}$.

Definition 3.6. *A **three-dimensional model** for the language* $\mathcal{L}$ *is a tuple* $\mathbb{M}_3 = (S \times T \times H, R_a^S, R^T, R^H, V_3)$, *where:*

(i) $\mathbb{M}_E = (S, R, V)$ *is an epistemic model,* $\mathbb{M}_T = (T, <, V)$ *is a temporal model, and* $\mathbb{M}_H = (H, \sqsubseteq, V)$ *is a historical model.*

---

[13]Where the converse relation of $<$ is $>$.
[14]Cf. [2].

(ii) $S \times T \times H = \{\langle s,t,h\rangle, \langle s',t',h'\rangle, \langle s'',t'',h''\rangle, \ldots\} = DOM(\mathbb{M}_3)$ *is the domain of* $\mathbb{M}_3$.

(iii) $R_a^S, R^T, R^H \subseteq (S \times T \times H)^2$ *are product accessibility relations defined as:*

- $\langle s,t,h\rangle \; R_a^S \; \langle s',t,h\rangle \quad$ iff $\quad sR_as'$
- $\langle s,t,h\rangle \; R^T \; \langle s,t',h\rangle \quad$ iff $\quad t < t'$
- $\langle s,t,h\rangle \; R^H \; \langle s,t,h'\rangle \quad$ iff $\quad h \sqsubseteq h'$

(iv) $V_3 : Var \to \mathcal{P}(S \times T \times H)$ *is a valuation function that assigns subsets of the domain to propositional variables.*

Product models, such as the three-dimensional model $\mathbb{M}_3$, provide a rigorous framework for integrating multiple modal dimensions—epistemic, temporal, and historical—within a single logical structure. These models allow for the simultaneous representation of knowledge, temporal progression, and historical branching, making them particularly useful in applications such as dynamic epistemic logic, artificial intelligence, and the formal analysis of historical narratives. The study of combined modal systems has been extensively developed in recent literature, with significant contributions on the interactions between different modal operators. A fundamental reference in this field is *Analysis and Synthesis of Logics* by Carnielli, Coniglio, and D'Ottaviano [2], which provides a systematic methodology for combining logics and analyzing their properties.

To fully utilize the expressive power of these models, we now introduce the notion of *satisfaction*, which formally defines how formulas of the language $\mathcal{L}$ are evaluated within a three-dimensional model.

Definition 3.7. *Let* $\langle s,t,h\rangle$ *be a three-dimensional state in the model* $\mathbb{M}_3$. *For every* $\varphi \in \mathcal{L}$, *satisfaction is defined as follows:*

- $\mathbb{M}_3, \langle s,t,h\rangle \models \varphi$ iff $\langle s,t,h\rangle \in V_3(\varphi)$

- $\mathbb{M}_3, \langle s,t,h\rangle \models \varphi \land \psi$ iff $\mathbb{M}_3, \langle s,t,h\rangle \models \varphi$ and $\mathbb{M}_3, \langle s,t,h\rangle \models \psi$

- $\mathbb{M}_3, \langle s,t,h\rangle \models K_a\varphi$ iff $\forall s' \in S$, if $\langle s,t,h\rangle R_a^S\langle s',t,h\rangle$ then $\mathbb{M}_3, \langle s',t,h\rangle \models \varphi$

- $\mathbb{M}_3, \langle s,t,h\rangle \models [F]\varphi$ iff $\forall t' \in T$, if $\langle s,t,h\rangle R_T\langle s,t',h\rangle$ then $\mathbb{M}_3, \langle s,t',h\rangle \models \varphi$

- $\mathbb{M}_3, \langle s,t,h\rangle \models [P]\varphi$ iff $\forall t' \in T$, if $\langle s,t,h\rangle \check{R}_T\langle s,t',h\rangle$ then $\mathbb{M}_3, \langle s,t',h\rangle \models \varphi$

- $\mathbb{M}_3, \langle s,t,h\rangle \models [H]\varphi$ iff $\forall h' \in H$, if $\langle s,t,h\rangle R_H\langle s,t,h'\rangle$ then $\mathbb{M}_3, \langle s,t,h'\rangle \models \varphi$

- $\mathbb{M}_3, \langle s, t, h \rangle \models K_a^\psi \varphi$ iff $\forall s' \in S$, $\exists t' \in T$, if $\langle s, t, h \rangle R_a^S \langle s', t, h \rangle$ then $\langle s', t, h \rangle R_T \langle s', t', h \rangle$ and if $\mathbb{M}_3, \langle s', t, h \rangle \models \psi$ then $\mathbb{M}_3, \langle s', t', h \rangle \models \varphi$

These satisfaction conditions define the interpretation of formulas in $\mathbb{M}_3$. The first clause states that a propositional variable φ holds at a state $\langle s, t, h \rangle$ if and only if it is assigned to that state by the valuation function V_3. The conjunction rule follows the standard interpretation of classical logic. The modal operator K_a encodes epistemic accessibility, meaning that an agent a knows φ if it holds in all epistemically accessible states in the domain S. Temporal modalities $[F]$ and $[P]$ represent future and past necessity, respectively, across the temporal dimension T. The operator $[H]$ governs hypothetical accessibility in H. Lastly, K_a^ψ introduces a knowledge-based temporal dependency, ensuring that if ψ holds at an epistemically accessible state, then φ holds in a future-related state.

Now that we have established the basic framework for satisfaction in three-dimensional models, it is time to turn our attention to the concept of logical consequence. A formula φ is said to be a logical consequence of a set of formulas Γ if φ holds in every model in which all the formulas in Γ hold. This is formalized as follows:

Definition 3.8 (Logical Consequence). *Let $\mathcal{L}$ is a tuple $\mathbb{M}_3 = (S \times T \times H, R_a^S, R^T, R^H, V_3)$ be a three-dimensional model and $\Gamma \subseteq \mathcal{L}$. We say that $\varphi \in \mathcal{L}$ is a logical consequence of Γ, denoted by $\Gamma \models \varphi$, if for every state $\langle s, t, h \rangle \in S \times T \times H$:*

$$\mathbb{M}_3, \langle s, t, h \rangle \models \psi \quad \text{for all} \quad \psi \in \Gamma \quad \text{implies} \quad \mathbb{M}_3, \langle s, t, h \rangle \models \varphi.$$

To conclude this section, we will analyze several classes of models that arise in this logic through the combination of specific properties from each type of model. We will demonstrate that it is possible to define hybrid model classes where temporal accessibility satisfies certain conditions, while epistemic accessibility meets others. Let us explore these concepts through the following definition.

Definition 3.9. *The classes of models that can be established for the three-dimensional system described depend on the properties of the accessibility relations (R_a, R_T, R_H) in the three components of the model. These classes are as follows:*

1. **Serial Model:** *The epistemic accessibility relation (R_a) is serial, meaning that for every state $s \in S$, there exists some state $s' \in S$ such that*

$\langle s,t,h\rangle R_a \langle s',t,h\rangle$. *The temporal relation (R_T) and historical relation (R_H) may or may not have additional properties. These models can belong to the class $\mathcal{KD}$ if the temporal relation is transitive and the historical relation is reflexive.*

2. **Reflexive Model:** *The epistemic accessibility relation (R_a) is reflexive, meaning that for every state $s \in S$, we have $\langle s,t,h\rangle R_a \langle s,t,h\rangle$. Depending on the properties of R_T and R_H, there may be several subclasses of reflexive models. These models can belong to the class $\mathcal{T}$ if the temporal relation is reflexive, or to a reflexive subclass of historical models if R_H is reflexive.*

3. **Transitive Model:** *The epistemic accessibility relation (R_a) is transitive, meaning that if $\langle s,t,h\rangle R_a \langle s',t,h\rangle$ and $\langle s',t,h\rangle R_a \langle s'',t,h\rangle$, then $\langle s,t,h\rangle R_a \langle s'',t,h\rangle$. If the temporal and historical relations are also transitive, the models can belong to the class $\mathcal{K}4$.*

4. **Symmetric Model:** *The epistemic accessibility relation (R_a) is symmetric, meaning that if $\langle s,t,h\rangle R_a \langle s',t,h\rangle$, then $\langle s',t,h\rangle R_a \langle s,t,h\rangle$. If the temporal relation is also symmetric, these models belong to the class $\mathcal{B}$.*

5. **Euclidean Model:** *The epistemic accessibility relation (R_a) is euclidean, meaning that if $\langle s,t,h\rangle R_a \langle s',t,h\rangle$ and $\langle s,t,h\rangle R_a \langle s'',t,h\rangle$, then $\langle s',t,h\rangle R_a \langle s'',t,h\rangle$. If the temporal and historical relations are also euclidean, these models belong to the class $\mathcal{K}45$.*

6. **Equivalence Model:** *If all three accessibility relations (R_a, R_T, and R_H) are equivalence relations, i.e., reflexive, symmetric, and transitive, the model is an equivalence model. These models belong to the class $\mathcal{S}5$.*

7. **Product Models:** *Product models combine different types of accessibility relations (epistemic, temporal, and historical) to represent the behavior of the involved logical systems. In this case, the combination of these relations determines a wide range of models that may be serial, reflexive, transitive, symmetric, euclidean, or equivalence, depending on how these relations interact.*

These model classes allow for the representation of a wide variety of scenarios in which the accessibility relations between epistemic, temporal, and historical states interact in different ways. This undoubtedly provides us with excellent material for future research. In Appendix 5, we present some relevant tautologies. For now, let us proceed with the cases that concern us, those of Ockham and Thomas Aquinas.

4 Some Models for Prophecies

In this section, we will formally analyze two kinds of models for prophecies. In the first subsection, we present a formal analysis of the Nineveh prophecy. In the second subsection, we provide a formal characterization of Aquinas' distinction between perfect prophecy and prophetic intuition.

4.1 Prophetic and Divine Knowledge in Nineveh Prophecy

As we saw above in Section 2.1, Ockham's proposal consists in considering prophecies as epistemic conditionals about the future. To address this, we introduce a new universal modality, called the *full omniscience modality*[15]:

Definition 4.1. $\mathbb{M}_3, \langle s, t, h \rangle \vDash K_\Delta \varphi$ iff $\forall s' \in S, t' \in T, h' \in H, \langle s, t, h \rangle\ R_\Delta\ \langle s', t', h' \rangle$ then $\mathbb{M}_3, \langle s', t', h' \rangle \vDash \varphi$

Considering the specific case of the prophecy of Nineveh, our formalization in the system is as follows:

$$\mathbb{M}_3, \langle s, t, h \rangle \vDash K_\pi^{-R} O$$

Where R stands for "the people of Nineveh repent" and O for "Nineveh is overthrown". Let us now examine how it is possible to maintain compatibility between human actions and divine knowledge. This formula represents the conditional knowledge of the prophet, denoted by π, which is imperfect. This prophecy serves to influence human agents, urging them to change their actions, namely to repent, so that the city is not overthrown. In this sense, the following formula is also valid within the same index:

$$\mathbb{M}_3, \langle s, t, h \rangle \vDash K_\pi^{-R} O \rightarrow R$$

It expresses that if the prophet knows that Nineveh will be overthrown unless the people repent, then their repentance is the condition that avoids the destruction of the city. In this case, the knowledge of the prophet is still conditional, as it depends on the human decision to repent. This reinforces the

---

[15]In Appendix 5, we will explore a possible formalization of a more general approach to this modality.

idea that the prophecy serves as a guide for changing human behavior to align with divine expectations.

The first formula represents the imperfect knowledge of the prophe[16], while the second reflects the free will of human beings in response to the prophecy. Furthermore, both expressions are consistent with the following formula:

$$\mathbb{M}_3, \langle s, t, h \rangle \vDash K_\Delta R$$

That is, God knows, at the same index, that the people of Nineveh repent. This is true not only in the present state, but for every index, since from the divine point of view, every state is identified.[17] Given the nature of the K_Δ modality, there is no incompatibility between the omniscient agent Δ accessing every state in every model and the agent's knowledge π implying R. The same information is accessible to different agents to varying degrees; on one hand, the omniscient agent accesses all possible information, while the prophet agent only accesses a portion of it. The main distinction lies in the fact that the information conveyed by prophecy is being considered from two perspectives. Traditionally, when discussing this problem, the question is posed solely from the human perspective, which seemingly creates an incompatibility between human action and God's perfect knowledge. However, from this perspective, it is possible to present the same information from both viewpoints.

This alleged problem is mitigated by considering that, from the perspective of the omniscient agent Δ, all states are identical. Thus, there is no distinction between past and future, nor between epistemically accessible information, nor between possible parallel histories. From our perspective, however, these indexes form a three-dimensional space, as we can only access information within our timeline, conditioned to receive partial knowledge and distinguishing between the past, present, and future.

From the divine perspective, nothing is inaccessible; every state of affairs, past, present, or future, is equally comprehensible and known. This omni-

---

[16]"Reply to Objection 1: The Lord reveals to the prophets all things that are necessary for the instruction of the faithful; yet not all to every one, but some to one, and some to another". [11, Q 171, Art. 4].

[17]"Reply to Objection 2: The Divine foreknowledge regards future things in two ways. First, as they are in themselves, in so far, to wit, as it sees them in their presentiality: secondly, as in their causes, inasmuch as it sees the order of causes in relation to their effects. And though future contingencies, considered as in themselves, are determinate to one thing, yet, considered as in their causes, they are not so determined but that they can happen otherwise. Again, though this twofold knowledge is always united in the Divine intellect, it is not always united in the prophetic revelation, because an imprint made by an active cause is not always on a par with the virtue of that cause". [11, Q 171, Art. 6].

science, which transcends the limitations of temporal and epistemic distinctions, offers a unique vantage point from which all truths are self-evident and immediate. Such a perspective, while difficult to fully capture within our finite understanding, can be approached through models that attempt to account for the infinite and all-encompassing nature of divine knowledge. In Appendix 5, we present an initial attempt to construct a logical framework that reflects this infinitary perspective, positing a universal agent whose knowledge spans all possible worlds and states, without constraint or limitation. This model represents a step towards a more comprehensive understanding of divine omniscience, though much remains to be explored in refining its formal structure and implications.

We will continue by analyzing models that capture Thomas Aquinas' distinction between perfect prophecy and prophetic intuition. This exploration will deepen our understanding of the nature and scope of divine knowledge as it pertains to human prophecy.

4.2 Perfect Prophecy and Prophetic Intuition

Thomas Aquinas distinguishes between two types of prophecy: *perfect prophecy* and *prophetic intuition*. These two forms of prophecy can be formalized using the concepts of introspection, both positive and negative, which have been refined and extended through the use of conditional knowledge operators. To better understand this distinction, we draw on the classical division between positive and negative introspection, representing one of these prophetic states within the framework of modal logic.

The distinction between perfect prophecy and prophetic intuition can be effectively captured using the conditional knowledge operators in our formal system. Perfect prophecy, analogous to positive introspection, is characterized by the following formula:

$$K_\pi^\psi \varphi \to K_\pi K_\pi^\psi \varphi$$

This formula suggests that if the prophet knows that a certain event will occur under a given condition ($K_\pi^\psi \varphi$), then the prophet knows that they know it ($K_\pi K_\pi^\psi \varphi$). In other words, perfect prophecy entails a complete and reflective knowledge of the future event, where the prophet's knowledge is self-aware and certain.

On the other hand, prophetic intuition, is captured by the following formula:

$$K_\pi^\psi \varphi \wedge \neg K_\pi K_\pi^\psi \varphi$$

This formula represents the situation where the prophet knows that a certain event will occur under a given condition ($K_\pi^\psi \varphi$), but does not have reflective knowledge of it ($\neg K_\pi K_\pi^\psi \varphi$). In this case, the prophet has knowledge of the future event, yet lacks certainty or awareness of their own knowledge regarding it. Prophetic intuition, therefore, involves a more limited or partial form of knowledge in comparison to perfect prophecy.

These differences highlight the philosophical tension between determinism and free will, as the nature of prophetic knowledge forces us to grapple with the limits of human understanding in relation to divine omniscience. The subsequent proofs demonstrate how these formulas are satisfied in the respective model classes, bridging the gap between philosophical intuition and formal logic. In this way, we establish a rigorous foundation for understanding prophecy through the lens of modal logic, with each model offering insights into the structure of divine and human knowledge, as well as the intricate relationship between them.

The models of prophecy explored—Ockham's epistemic conditional model and Thomas Aquinas' distinction between perfect prophecy and prophetic intuition—offer distinct yet complementary approaches to understanding the nature of prophetic knowledge. Ockham's model, grounded in epistemic conditions, highlights the relationship between divine omniscience and human freedom, while Aquinas' distinction emphasizes the varying degrees of prophetic certainty. Together, these models provide a comprehensive framework for analyzing the logical structure of prophecy, allowing for a nuanced understanding of how divine and human knowledge intersect within the bounds of logical and philosophical reasoning. Now we take a closer look at the theorems corresponding to perfect prophecy and prophetic intuition.

Theorem 4.1 (Perfect Prophecy Theorem). *Let $\mathbb{M}_3 = (S \times T \times H, R_a^S, R^T, R^H, V_3)$ be a three-dimensional model, and let π be a prophet. Then, the following formula holds:*

$$K_\pi^\psi \varphi \to K_\pi K_\pi^\psi \varphi$$

Proof. We will prove that if $K_\pi^\psi \varphi$ holds in a world, then $K_\pi K_\pi^\psi \varphi$ also holds in that world, using the conditions of the three-dimensional model.

- **Assumption:** Assume $\mathbb{M}_3, \langle s, t, h \rangle \models K_\pi^\psi \varphi$, meaning that the prophet π knows φ under the condition of his own knowledge, i.e., φ holds in all worlds that are accessible from $\langle s, t, h \rangle$ by the relation $\mathcal{R}_\pi^\psi$.

- **Reflexivity of K_π:** Since $\mathbb{M}_3$ is a reflexive model, it satisfies $\mathcal{R}_\pi$ for all worlds. This means that if the prophet π knows φ, then he also knows that he knows φ. Formally, $\mathbb{M}_3, \langle s, t, h \rangle \models K_\pi K_\pi^\psi \varphi$.

- **Conclusion:** Therefore, we have shown that $K_\pi^\psi \varphi$ implies $K_\pi K_\pi^\psi \varphi$, completing the proof of the Perfect Prophecy Theorem.

$\square$

Theorem 4.2 (Prophetic Intuition Theorem). *Let $\mathbb{M}_3 = (S \times T \times H, R_a^S, R^T, R^H, V_3)$ be a three-dimensional model, and let π be a prophet. Then, the following formula holds:*

$$K_\pi^\psi \varphi \wedge \neg K_\pi K_\pi^\psi \varphi$$

Proof. We will prove that if $K_\pi^\psi \varphi$ holds in a world, but $K_\pi K_\pi^\psi \varphi$ does not hold, then the formula holds in non-reflexive models.

- **Assumption:** Assume $\mathbb{M}_3, \langle s, t, h \rangle \models K_\pi^\psi \varphi$, meaning that the prophet π knows φ under the condition of his own knowledge, i.e., φ holds in all worlds that are accessible from $\langle s, t, h \rangle$ by the relation $\mathcal{R}_\pi^\psi$.

- **Non-reflexivity of K_π:** We now need to show that $K_\pi K_\pi^\psi \varphi$ does not hold. This is the case when the model $\mathbb{M}_3$ is non-reflexive for $\mathcal{R}_\pi$. Specifically, there exists some world $\langle s', t', h' \rangle$ where the prophet does not know that he knows φ, i.e., $\mathbb{M}_3, \langle s', t', h' \rangle \not\models K_\pi K_\pi^\psi \varphi$.

- **Conclusion:** Therefore, we have shown that $K_\pi^\psi \varphi$ and $\neg K_\pi K_\pi^\psi \varphi$ can hold together in non-reflexive models, completing the proof of the Prophetic Intuition Theorem.

$\square$

The formulas that correspond to these types of prophecy provide a clear contrast in terms of the accessibility relations between states in our models. As we have seen, perfect prophecy implies a model that is reflexive and transitive, where the prophet's knowledge not only encompasses the future event but also extends to their knowledge of that knowledge. In contrast, prophetic intuition represents a more nuanced scenario, where the prophet may possess knowledge of an event without the epistemic self-awareness of their own knowledge, which is captured by models that are partially reflexive and not fully transitive.

5   Conclusion

In this work, we have navigated the complex landscape of prophecy, focusing on the distinction between perfect prophecy and prophetic intuition as formulated by Thomas Aquinas, while also integrating Ockham's epistemic approach to prophecy. Through formalizing these concepts within the framework of conditional knowledge and the interplay between human free will and divine omniscience, we have established a deeper understanding of how prophecies operate within logical systems. By comparing and contrasting Aquinas' views with Ockham's model, we have illustrated that while the former emphasizes the clarity of divine knowledge and its accessibility to the prophet, the latter highlights the conditional nature of prophecies, where knowledge is partial and future events remain contingent on human action. This distinction between perfect prophecy and prophetic intuition reflects a broader philosophical tension between determinism and freedom, knowledge and uncertainty, which we have formalized through logical semantics.

Our exploration has led to a nuanced analysis of the models for both types of prophecy, shedding light on how these models interact within different classes of models based on their satisfaction conditions. The logical formulations presented, alongside the corresponding proofs, demonstrate how the perfect knowledge of the omniscient agent aligns with the prophetic knowledge in Aquinas' perfect prophecy, while in Ockham's model, prophetic intuition operates within a framework of human decision-making and divine foreknowledge. In conclusion, this study not only provides a formal logical understanding of prophetic knowledge but also opens avenues for further exploration of divine omniscience, free will, and the nature of future knowledge in philosophical and theological discourse.

Appendix 1

Now, in this appendix, we present a list of some valid formulas in this system.

Theorem 5.1. $\mathbb{M}_3, \langle s, t, h \rangle \models K_a \varphi \rightarrow K_a^\psi \varphi$

Proof. Assume that
$$\mathbb{M}_3, \langle s, t, h \rangle \models K_a \varphi.$$
By definition of the epistemic operator K_a, for every state s' with
$$\langle s, t, h \rangle R_a^S \langle s', t, h \rangle,$$

we have
$$\mathbb{M}_3, \langle s', t, h \rangle \vDash \varphi.$$

Hence, in any such state where the condition ψ is (or might be) assumed to hold, φ is already true. Thus, by the definition of the conditional knowledge operator K_a^ψ,
$$\mathbb{M}_3, \langle s, t, h \rangle \vDash K_a^\psi \varphi.$$

Therefore,
$$K_a \varphi \to K_a^\psi \varphi$$
holds in $\mathbb{M}_3, \langle s, t, h \rangle$. □

Theorem 5.2. $\mathbb{M}_3, \langle s, t, h \rangle \vDash K_a^\psi \varphi \to K_a K_a^\psi \varphi$

Proof. Assume
$$\mathbb{M}_3, \langle s, t, h \rangle \vDash K_a^\psi \varphi.$$

This means that for every s' with
$$\langle s, t, h \rangle \, R_a^S \, \langle s', t, h \rangle,$$
if the condition ψ holds then, after some appropriate temporal transition, φ holds. Consequently, in every state accessible from $\langle s, t, h \rangle$ the agent also has the conditional knowledge $K_a^\psi \varphi$. That is,
$$\mathbb{M}_3, \langle s, t, h \rangle \vDash K_a K_a^\psi \varphi.$$

Hence, the tautology is established. □

Theorem 5.3. $\mathbb{M}_3, \langle s, t, h \rangle \vDash K_a^\psi \varphi \to \neg K_a \neg K_a^\psi \varphi$

Proof. Assume
$$\mathbb{M}_3, \langle s, t, h \rangle \vDash K_a^\psi \varphi.$$

Then, by definition, the agent knows conditionally that φ holds under ψ in all epistemically accessible states. It is therefore impossible that the agent knows the negation of this conditional knowledge. Thus,
$$\mathbb{M}_3, \langle s, t, h \rangle \vDash \neg K_a \neg K_a^\psi \varphi.$$

This proves the tautology. □

Theorem 5.4. $\mathbb{M}_3, \langle s, t, h \rangle \vDash K_a^\psi \varphi \to K_a(\psi \to \langle F \rangle \varphi)$

Proof. Assume
$$\mathbb{M}_3, \langle s, t, h \rangle \models K_a^\psi \varphi.$$

This means that in every epistemically accessible state s', if ψ holds then, after some future transition (via R_T), φ holds. Thus, the agent knows that whenever ψ is true, eventually φ will be true. Formally, this is exactly the meaning of
$$K_a(\psi \to \langle F \rangle \varphi).$$

Hence, the tautology holds. □

Theorem 5.5. $\mathbb{M}_3, \langle s, t, h \rangle \models K_a^\psi \varphi \to \left([H]\langle F \rangle K_a^\psi \to \langle F \rangle \varphi \right)$

Proof. Assume
$$\mathbb{M}_3, \langle s, t, h \rangle \models K_a^\psi \varphi.$$

Now, suppose that
$$\mathbb{M}_3, \langle s, t, h \rangle \models [H]\langle F \rangle K_a^\psi.$$

This means that in every historical branch, in every future accessible state, the agent conditionally knows φ under ψ. Since by assumption $K_a^\psi \varphi$ holds, it follows that in all future states, φ must eventually hold. Therefore,
$$\mathbb{M}_3, \langle s, t, h \rangle \models \langle F \rangle \varphi.$$

Hence, the tautology is established. □

Theorem 5.6. $\mathbb{M}_3, \langle s, t, h \rangle \models [H]K_a^\psi \varphi \to K_a\left([H]\psi \to [H]\varphi\right)$

Proof. Assume
$$\mathbb{M}_3, \langle s, t, h \rangle \models [H]K_a^\psi \varphi.$$

Then, in every historical branch, the agent has conditional knowledge that φ holds under ψ. Hence, in every epistemically accessible state, if ψ holds throughout the history (i.e., $[H]\psi$), then φ holds throughout the history (i.e., $[H]\varphi$). This is precisely the content of
$$K_a([H]\psi \to [H]\varphi),$$

proving the tautology. □

Theorem 5.7. $\mathbb{M}_3, \langle s, t, h \rangle \models K_a^\perp \varphi$

Proof. Here, $K_a^\perp \varphi$ denotes that the agent knows φ under an impossible condition (denoted by $\perp$, falsehood). Since $\perp$ is never true, the condition is vacuously satisfied, and thus $K_a^\perp \varphi$ holds trivially. $\square$

Theorem 5.8. $\mathbb{M}_3, \langle s, t, h \rangle \vDash K_a^\psi \top$

Proof. Since $\top$ is always true, regardless of the condition ψ, the agent trivially has conditional knowledge of $\top$. Therefore, $K_a^\psi \top$ holds in $\mathbb{M}_3, \langle s, t, h \rangle$. $\square$

Appendix 2: Omega-Dimensional Models for Divine Knowledge

In this appendix, we introduce the formalization of *omega-dimensional models* that attempt to capture divine knowledge, expanding beyond the standard three-dimensional models. These models extend to a transfinite number of dimensions, represented by the ordinal set ω, to reflect the multiple modalities of divine knowledge.

Definition of Omega-Dimensional Models

An *omega-dimensional model* $M = \langle W, \{R_\alpha\}_{\alpha \in \omega}, V, \{K^\alpha\}_{\alpha \in \omega} \rangle$ is defined as a structure such that:

- W is a non-empty set of *worlds* or *possible states*.

- $\{R_\alpha\}_{\alpha \in \omega}$ is a family of accessibility relations $R_\alpha \subseteq W \times W$ for each ordinal $\alpha \in \omega$, where each R_α represents an *accessibility relation of dimension* α. These relations correspond to different kinds of connections between worlds, such as knowledge, time, counterfactuals, etc.

- V is a *truth function* that assigns a truth value to each proposition in each world $w \in W$ and in each dimension α, i.e., $V : W \times P \to \{\text{True}, \text{False}\}$, where P is the set of propositions.

- $\{K^\alpha\}_{\alpha \in \omega}$ is a set of modal operators K^α, one for each dimension α, which determines the agent's knowledge of the world from the perspective of the dimension α.

Satisfaction in Omega-Dimensional Models

The satisfaction relation ⊨ in the context of an omega-dimensional model is recursive, depending on the accessibility relations R_α in each dimension α. Formally:

- $M, w \models \varphi$ if φ is true in world w according to the truth function V.

- $M, w \models K^\alpha \varphi$ if $\forall w' \in R_\alpha(w), M, w' \models \varphi$. This means that φ is true in all worlds accessible from w under the accessibility relation of dimension α.

- $M, w \models K^\Omega \varphi$ if $\forall w' \in R_\Omega(w), M, w' \models \varphi$, where Ω is the ordinal representing omniscience (absolute access to all dimensions).

Divine Omniscience and Transfinite Knowledge

The operator of *divine omniscience* O_a^Ω can be defined as follows:

$$M, w \models O_a^\Omega \varphi \iff \forall \alpha \in \omega, \forall w' \in R_\alpha(w), M, w' \models \varphi$$

This means that divine knowledge O_a^Ω extends over all dimensions, from temporal and epistemic dimensions to ontological dimensions, ensuring that omniscience encompasses all realities accessible from any world.

Modalities of Divine Knowledge in an Omega-Dimensional Model

We can also define the *universal accessibility relation* in terms of ordinals:

- **Total knowledge of all accessible worlds**:

  $$M, w \models \forall \alpha \in \omega, K^\alpha \varphi \iff \forall \alpha \in \omega, \forall w' \in R_\alpha(w), M, w' \models \varphi$$

  This means that, in every dimension α, the agent knows all worlds accessible from w under the relation R_α.

- **Perfect knowledge (omniscience)**:

  $$M, w \models O_a^\Omega \varphi \iff \varphi \text{ is true in all worlds of all dimensions}$$

  This expresses the idea that divine knowledge spans all dimensions of reality.

Consequences of Omega-Dimensional Models

Divine omnipresence: In these models, divine omniscience is not limited to a specific dimension (such as time or space), but extends over a transfinite number of dimensions. This allows the modeling of a divine entity whose knowledge encompasses the entire structure of reality, not just the facts of the present world, but also counterfactuals, possible futures, and alternative realities.

Access to all realities: A divine agent is not limited to what is accessible within the confines of a particular dimension; rather, they have full access to all possible worlds within any dimension of their reality.

References

[1] *Holy Bible: New Revised Standard Version*. National Council of Churches of Christ in the USA, New York, 1989. A widely accepted academic translation, used in theological and philosophical research.

[2] CARNIELLI, W., CONIGLIO, M., GABBAY, D. M., GOUVEIA, P., AND SERNADAS, C. *Analysis and Synthesis of Logics: how to cut and paste reasoning systems*, vol. 35. Springer Science & Business Media, 2008.

[3] CRUZ, J. D. G., AND RAMOS, Y. E. Opposition relations between prophecies. In *International Conference on Theory and Application of Diagrams* (2020), Springer, pp. 394–401.

[4] DE OCKHAM, G. *De praedestinatione et de praescientia Dei respectu futurorum contingentium*, vol. 5 of *Opera Philosophica et Theologica*. Franciscan Institute Publications, St. Bonaventure, NY, 1983. Edición crítica de los escritos de Guillermo de Ockham.

[5] EDIDIN, A., AND NORMORE, C. Oackham on prophecy. *International Journal for Philosophy of Religion* (1982), 179–189.

[6] GORANKO, V. *Temporal Logics*. Cambridge University Press, 2023.

[7] KURUCZ, A., WOLTER, F., ZAKHARYASCHEV, M., AND GABBAY, D. M. *Many-dimensional modal logics: Theory and applications*, vol. 148. Elsevier, 2003.

[8] LIMONTA, R., AND FEDRIGA, R. Assensum in mente prophetae: William of ockham and walter chatton on prophecies. *Analiza i Egzystencja 54* (2021), 57–80.

[9] MAINZER, K., AND CENTRONE, S. *Temporal Logic: From Philosophy and Proof Theory to Artificial Intelligence and Quantum Computing*. World Scientific, 2023.

[10] MARX, M., VENEMA, Y., MARX, M., AND VENEMA, Y. *Multi-dimensional modal logic*. Springer, 1997.

[11] NICOLAI, J. S. *Thomae Aquinatis summa theologica: De justitia. De fortitudine. De temperantia. De prophetia. De quibusdam speciatim ad homines certae alicujus conditionis pertinentibus*. No. v. 5 in S. Thomae Aquinatis summa theologica: De

justitia. De fortitudine. De temperantia. De prophetia. De quibusdam speciatim ad homines certae alicujus conditionis pertinentibus. Guerin, 1870.

[12] OF OCKHAM, W. *Predestination, God's Foreknowledge, and Future Contingents*. Hackett Publishing Company, Indianapolis, 1983. Translation of: Tractatus de praedestinatione et de praescientia Dei et de futuris contingentibus.

[13] ØHRSTRØM, P., AND HASLE, P. *Temporal logic: From ancient ideas to artificial intelligence*, vol. 57. Springer Science & Business Media, 2007.

[14] SOLER, C. R. Ockham on the puzzle of prophecy and future contingency. *Journal of the History of Philosophy 62*, 4 (2024), 567–592.

[15] THOMASON, R. H. Combinations of tense and modality. In *Handbook of Philosophical Logic: Volume II: Extensions of Classical Logic*. Springer, 1984, pp. 135–165.

[16] ZANARDO, A. Branching-time logic with quantification over branches: The point of view of modal logic. *The Journal of Symbolic Logic 61*, 1 (1996), 1–39.

[17] ZANARDO, A., AND CARMO, J. Ockhamist computational logic: Past-sensitive necessitation in ctl. *Journal of Logic and Computation 3*, 3 (1993), 249–268.

Towards Non-Deductive Term Functor Logic

J.-Martín Castro-Manzano

1 Introduction

Term Functor Logic ($\mathcal{TFL}$, for short) is a relatively novel logic that follows the tradition of Aristotelian logic —hence its alternative name, "Traditional Formal Logic"— in the sense that it uses a term syntax rather than a Fregean syntax; however, it still needs some tweaks in order to claim its rightful place within the realm of Aristotelian logic —not that it needs to, but we would like to see it there! For instance, in other places we have offered some ways with which we can update $\mathcal{TFL}$ in order to comply with some criteria for relevance insofar as Aristotelian logic requires some sort of relevance [1]. And so, following this train of thought, in this contribution we try to update $\mathcal{TFL}$ by adding some ways with which we can deal with non-deductive inference, namely, inductive and abductive inference, insofar as Aristotelian logic demands the treatment of non-deductive inference. Thus, in order to reach this goal, in this contribution we start to combine $\mathcal{TFL}$ and a proxy of Non-Axiomatic Logic —which is a term logic that deals with non-deductive inference by desing— by way of a tableaux method. The result is a prototype of a tableaux method within the framework of $\mathcal{TFL}$ that would be able to model non-deductive inference.

2 Preliminaries

2.1 Deduction, Induction, and Abduction

Against popular opinion, Aristotelian logic *is not* syllogistic: it is more. Aristotelian logic includes theories about predication, statements, inferences, explanations, argumentation, and inferential mistakes. Syllogistic is only a fragment of Aristotelian logic, it is a theory of deductive inference crafted to avoid irrelevance; but Aristotelian logic is more than that. Aristotelian logic demands the ability to point out conclusions that cannot be necessarilly proven but are possible, and the skill to indicate instances in which non-deductive patterns may be useful [6]. Indeed, as pointed out by Peirce (CP 2.619), Aristotelian logic takes into account both deductive and non-deductive inference patterns, namely, inductive and abductive inference patterns.

Following Wang [10], we can represent these patterns, in a traditional manner, by paying attention to the position of the middle term, M, as in Table 1. Intuitively, deduction gives us *proof* that S is P via the transitity provided by the middle term M; induction gives us a *chance* that S is P via the common cause M; and abduction gives us *probable cause* that S is P via the common effect M. These intuitive, informal notions can be better understood with NARS.

Deduction	Induction	Abduction
1. M is P	1. M is P	1. P is M
2. S is M	2. M is S	2. S is M
$\therefore S$ is P	$\therefore S$ is P	$\therefore S$ is P

Table 1: Deduction, induction, and abduction in tradition.

2.2 NARS

NARS (short for Non-Axiomatic Reasoning System) is an artificial reasoning system designed after Non-Axiomatic Logic [10], a logic of terms defined in virtue of some Aristotelian insights. In NARS, a statement has the form $S \subset P \langle F, C \rangle$ where S is the subject term of the statement, and P is the predicate term so that this statement says that "S is P" in English; and $\langle F, C \rangle$ is a pair of real numbers, $F, C \in [0, 1]$, referred to as the *frequency* and the *confidence* of the statement, respectively. So, for example, the statement

$$Tweety \subset Bird \langle 1, .8 \rangle$$

is saying that "Tweety is a bird" is true with a high degree of confidence.

Given this definition of a statement, the informal patterns of deduction, induction, and abduction previously deployed can be formalized in NARS as in Table 2.

Deduction, we believe, requires no examples; but induction and abduction surely do. Hence suppose, for instance, that "Tweety is a bird" with a high degree of confidence, and that "Twetty flies" has a balanced degree of confidence:

Deduction	Induction	Abduction
1. $M \subset P \langle F_1, C_1 \rangle$	1. $M \subset P \langle F_1, C_1 \rangle$	1. $P \subset M \langle F_1, C_1 \rangle$
2. $S \subset M \langle F_2, C_2 \rangle$	2. $M \subset S \langle F_2, C_2 \rangle$	2. $S \subset M \langle F_2, C_2 \rangle$
$\therefore S \subset P \langle F_d, C_d \rangle$	$\therefore S \subset P \langle F_i, C_i \rangle$	$\therefore S \subset P \langle F_a, C_a \rangle$
where	where	where
$F_d = F_1 F_2/(F_1 + F_2 - F_1 F_2)$	$F_i = F_1$	$F_a = F_2$
$C_d = C_1 C_2 (F_1 + F_2 - F_1 F_2)$	$C_i = F_2 C_1 C_2 / (F_2 C_1 C_2 + 1)$	$C_a = F_1 C_1 C_2 / (F_1 C_1 C_2 + 1)$

Table 2: Deduction, induction, and abduction in NARS [10].

$$Tweety \subset Bird \langle 1, .8 \rangle$$

$$Tweety \subset Flies \langle 1, .5 \rangle$$

then we can conclude, by way of induction, that

$$Flies \subset Bird \langle 1, .2 \rangle$$

that is to say, that "that which flies is a bird" is true with a low degree of confidence. And, for example, if we start with "If it rains then the floor is wet" and "Today the floor is wet" both with a high degree of confidence:

$$Rain \subset Wet \langle 1, .9 \rangle$$

$$Today \subset Wet \langle 1, .9 \rangle$$

then we can conclude, by way of abduction, that

$$Today \subset Rain \langle 1, .4 \rangle$$

that is to say, that "Today has rained" with less than a half degree of confidence. We will return to these concepts later.

2.3 $\mathcal{TFL}$

Term Functor Logic [8, 9, 3, 4, 5] is a plus-minus algebra that employs terms and functors, in Aristotelian fashion, rather than first order language elements such as individual variables or quantifiers. According to this algebra, the four categorical statements of syllogistic can be represented by the following syntax [4]:

$$\text{All S is P} := -S + P$$

$$\text{All S is not P} := -S - P$$
$$\text{Some S is P} := +S + P$$
$$\text{Some S is not P} := +S - P$$

Given this representation, $\mathcal{TFL}$ provides a simple rule for syllogistic inference: a conclusion follows validly from a set of premises if and only if *i)* the sum of the premises is algebraically equal to the conclusion and *ii)* the number of conclusions with particular quantity (*viz.*, zero or one) is the same as the number of premises with particular quantity [4, p.167]. Thus, for instance, if we consider a valid syllogism we can see how the application of this rule produces the right conclusion (Table 3).

Statement	$\mathcal{TFL}$
1. All computer scientists are animals.	$-C + A$
2. All logicians are computer scientists.	$-L + C$
⊢ All logicians are animals.	$-L + A$

Table 3: A valid syllogism.

In this example, we can clearly see how the rule works: *i)* if we add up the premises we obtain the algebraic expression $(-C + A) + (-L + C) = -C + A - L + C = -L + A$, so that the sum of the premises is algebraically equal to the conclusion and the conclusion is of the form $-L + A$, rather than $+A - L$, because *ii)* the number of conclusions with particular quantity (zero in this case) is the same as the number of premises with particular quantity (zero in this case).[1] In contrast, just for the sake of comparison, consider an invalid syllogism that does not add up (Table 4).

Statement	$\mathcal{TFL}$
1. All computer scientists are animals.	$-C + A$
2. All computer scientists are logicians.	$-C + L$
⊬ All logicians are animals.	$-L + A$

Table 4: An invalid syllogism.

Now, as exposed elsewhere and following [2, 7], we can develop a tableaux proof method for $\mathcal{TFL}$. So, let us say a *tableau* is an acyclic connected graph

---

[1] Although we are exemplifying this logic with syllogistic inferences, this system is capable of representing relational, singular, and compound inferences with ease and clarity. Furthermore, $\mathcal{TFL}$ is arguably more expressive than classical first order logic [3, p.172].

determined by nodes and vertices. The node at the top is called *root*. The nodes at the bottom are called *tips*. Any path from the root down a series of vertices is a *branch*. To test an inference for validity we construct a tableau which begins with a single branch at whose nodes occur the premises and the rejection of the conclusion: this is the *initial list*. We then apply the expansion rules that allow us to extend the initial list (Figure 1).

Figure 1: $\mathcal{TFL}$ tableaux expansion rules.

Figure 1a depicts the rule for universal statements, while Figure 1b shows the rule for particular statements. After applying a rule we introduce some index $i \in \{1, 2, 3, \ldots\}$. For universal statements the index may be any natural number; for particular statements the index has to be a fresh natural if they do not already have an index. Also, following $\mathcal{TFL}$ tenets, we assume the following rules of rejection: $-(\pm T) = \mp T$, $-(\pm T \pm T) = \mp T \mp T$, and $-(--T--T) = +(-T)+(-T)$.

A tableau is *complete* if and only if every rule that can be applied has been applied. A branch is *closed* if and only if there are terms of the form $\pm T^i$ and $\mp T^i$ on two of its nodes; otherwise it is *open*. A closed branch is indicated by writing a $\bot$ at the end of it; an open branch is indicated by writing ∞. A tableau is *closed* if and only if every branch is closed; otherwise it is *open*. So, as usual, $\pm T$ is a logical consequence of the set of terms Γ (i.e. $\Gamma \vdash \pm T$) if and only if there is a complete closed tableau whose initial list includes the terms of Γ and the rejection of $\pm T$ (i.e. $\Gamma \cup \{\mp T\} \vdash \bot$). As an example, consider Figure 2, which shows the inferences exposed in Tables 3 and 4.

3 Non-Deductive Tableaux

After these preliminaries, three facts should be clear: *i)* deduction, induction, and abduction are different forms of reasoning whose patterns can be distinguished by the position of a middle term given the framework of a logic of terms (by tradition); *ii)* there is a logic of terms in which we can distinguish

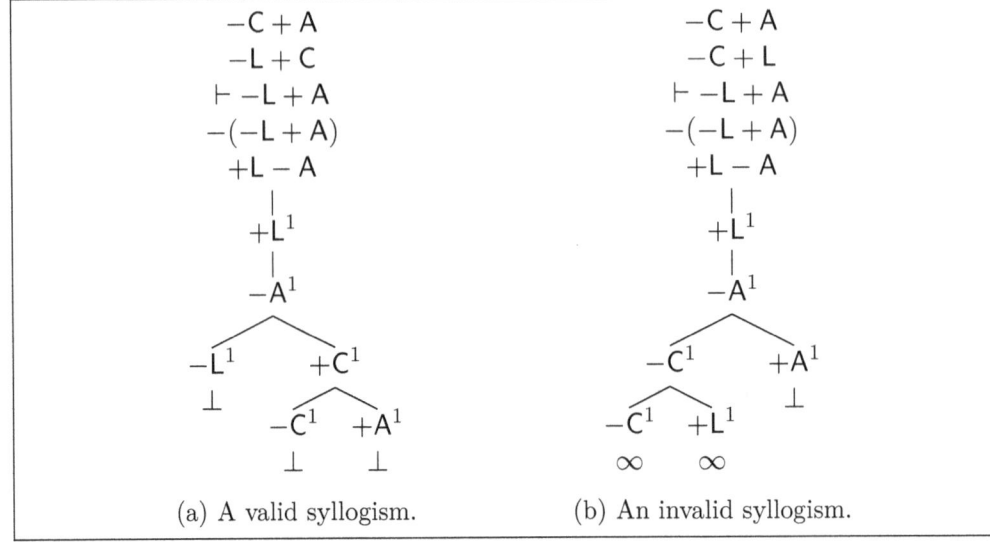

Figure 2: A pair of examples.

and compute deduction, induction, and abduction (by NARS); and *iii)* there is a tableaux proof method for a logic of terms that deals with deduction (by $\mathcal{TFL}$). Let us put these facts to work.

We will develop a $\mathcal{TFL}$ tableaux prototype to model deduction, induction, and abduction by paying attention to the position of the terms in a given syllogism. First we will translate the NARS statements into $\mathcal{TFL}$ statements, then we will offer some rules, and last we will put the proposal to the test. We will refer to this prototype as $\mathcal{TFL}^{DIA}$ for "$\mathcal{TFL}$ coupled with deduction, induction, and abduction."

3.1 Translation

As we mentioned above, in NARS a statement has the form $S \subset P\langle F, C\rangle$ and can be read as "S is P" in English; but there is some ambiguity though, for how should we interpret such a statement: as a universal, singular, or particular statement, or rather as a conditional statement? Consider the following examples:

$$Flies \subset Bird\langle 1, .2\rangle$$

$$Tweety \subset Bird\langle 1, .8\rangle$$

$$Rain \subset Wet\langle 1, .9\rangle$$

The first example can be understood as a universal statement (i.e. "That which flies is a bird"), the second one is clearly singular (i.e. "Tweety is a bird"), and the third one can be read as a conditional (i.e. "If it rains then the floor is wet"). What should we make of this? We think a $\mathcal{TFL}$ rendition of these statements should behave simply as a $-+$ combination given that the meaning of universal, singular, and conditional statements can be represented by $-+$ combinations in $\mathcal{TFL}$. Hence, our proposal will work with such combinations only and so the previous statements would look like this within $\mathcal{TFL}$:

$$-F + B\langle 1, .2\rangle$$
$$-T + B\langle 1, .8\rangle$$
$$-R + W\langle 1, .9\rangle$$

And in order to reduce cluttering, since we will be using indices when working with the tableaux, let us send the frequency and confidence values down, as follows:

$$-F + B_{\langle 1,.2\rangle}$$
$$-T + B_{\langle 1,.8\rangle}$$
$$-R + W_{\langle 1,.9\rangle}$$

However, this proposal is not complete, for we lack some way to represent particular statements. Now, one advantage of using $\mathcal{TFL}$ is that we can add the presence of particular statements to avoid ambiguity, for suppose we want to say that "Some bird flies" with a very high degree of confidence, how should we represent this statement? Like so:

$$+B + F_{\langle 1,.9\rangle}$$

And thus the general syntax for TFL^{DIA} would be $\pm S \pm P_{\langle F,C\rangle}$.

3.2 Rules

With this representation, our expansion rules look as follows:

Then, given any tableau, we say a branch is *closed* if and only if there are terms of the form $\pm T^i$ and $\mp T^i$ on it; otherwise, it is not-closed. A not-closed branch comes in various forms: a branch is *semi-closed* if and only if it has negative middle terms or positive subject terms, that is to say, if it has terms of the form $-M^i$ and $-M^i$, or $+S^i$ and $+S^i$ on it; a branch is *semi-open* if and

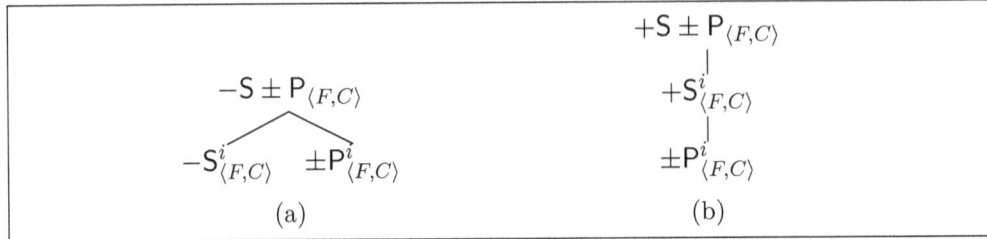

Figure 3: TFL^{DIA} expansion rules.

only if it has positive middle terms or negative predicate terms, that is to say, if it has terms of the form +M^i and +M$^{\tilde{i}}$, or −P^i and −P$^{\tilde{i}}$ on it. Otherwise, the branch is *open*. An open branch is indicated by writing ∞ at the end of it; a semi-closed (resp. semi-open) branch is indicated by writing ∝ (resp. ∞); and a closed branch is denoted by ⊥. We say, then, that a tableau is closed if and only if every branch is closed; a tableau is semi-closed if and only if it has closed and semi-closed branches; a tableau is semi-open if and only if it has closed and semi-open branches; otherwise, the tableau is open.

Given these remarks, we say ±T$_{\langle F,C \rangle}$ is a deductive consequence of the set of terms Γ (i.e. Γ ⊢$_D$ ±T$_{\langle F,C \rangle}$) if and only if there is a complete closed tableau whose initial list includes the terms of Γ and the rejection of ±T. We say that ±T$_{\langle F,C \rangle}$ is an inductive consequence of the set of terms Γ (i.e. Γ ⊢$_I$ ±T$_{\langle F,C \rangle}$) if and only if there is a complete semi-closed tableau whose initial list includes the terms of Γ and the rejection of ±T. And last, we say that ±T$_{\langle F,C \rangle}$ is an abductive consequence of the set of terms Γ (i.e. Γ ⊢$_A$ ±T$_{\langle F,C \rangle}$) if and only if there is a complete semi-open tableau whose initial list includes the terms of Γ and the rejection of ±T (Figure 4).

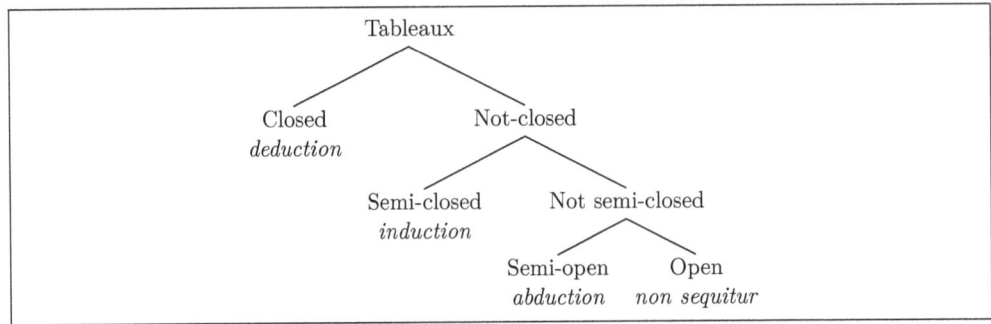

Figure 4: $\mathcal{TFL}^{DIA}$ types of inference.

Informally, with respect to the position of the middle terms: deductive trees have all their branches closed; inductive trees have their negative middle and subject terms branches open, but their predicate branches closed; and abductive trees have their positive middle and predicate terms branches open, but their subject branches closed.

3.3 Examples

To put this prototype to the test, let us review some examples. Consider the inductive and abductive inferences shown in section 2.2.

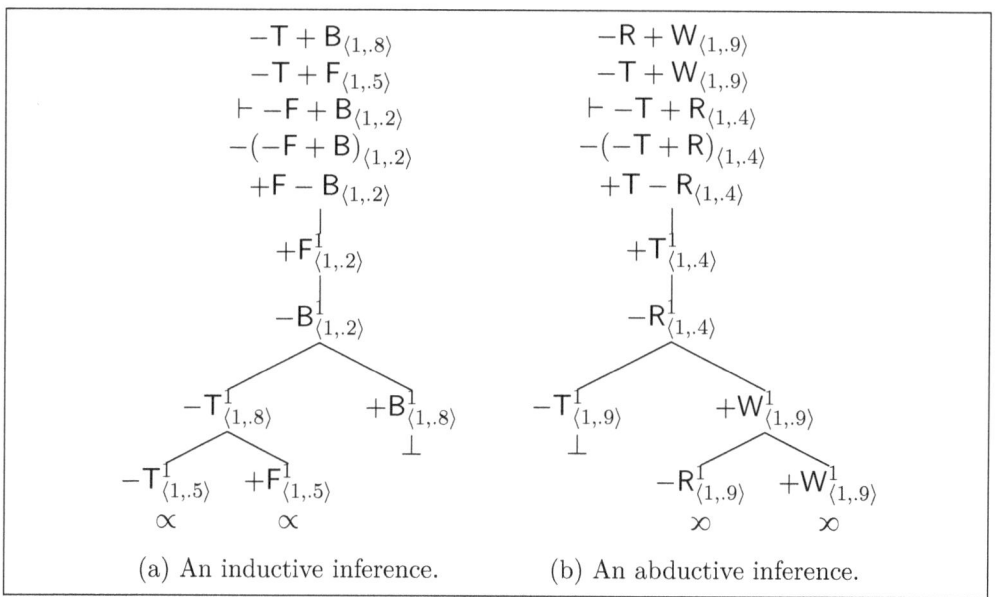

Figure 5: A pair of examples.

4 Final Remarks

In this contribution we have combined $\mathcal{TFL}$ and NARS by way of a tableaux method, and the result is a prototype for a tableaux method capable of dealing with non-deductive inference in a terministic, Aristotelian fashion. Our future work consists in including the frequency and confidence values, for instance, in terms of trichotomy (Table 5).

	Deduction	Induction	Abduction
Tableaux	Closed	Semi-closed	Semi-open
Values	=	>	<

Table 5: Deduction, induction, and abduction in $\mathcal{TFL}^{DIA}$.

References

[1] CASTRO-MANZANO, J.-M. Syllogistic relevance and term logic. *Journal of Logic, Language and Information 33*, 2 (Aug. 2024), 89–105.

[2] D'AGOSTINO, M., GABBAY, D. M., HÄHNLE, R., AND POSEGGA, J. *Handbook of Tableau Methods*. Springer, 1999.

[3] ENGLEBRETSEN, G. *The New Syllogistic*. Peter Lang, 1987.

[4] ENGLEBRETSEN, G. *Something to Reckon with: The Logic of Terms*. University of Ottawa Press, 1996.

[5] ENGLEBRETSEN, G., AND SAYWARD, C. *Philosophical Logic: An Introduction to Advanced Topics*. Bloomsbury Academic, 2011.

[6] MOZES, E. A deductive database based on aristotelian logic. *Journal of Symbolic Computation 7*, 5 (1989), 487 – 507.

[7] PRIEST, G. *An Introduction to Non-Classical Logic: From If to Is*. Cambridge Introductions to Philosophy. Cambridge University Press, 2008.

[8] SOMMERS, F. *The Logic of Natural Language*. Clarendon Press; Oxford: New York: Oxford University Press, 1982.

[9] SOMMERS, F., AND ENGLEBRETSEN, G. *An Invitation to Formal Reasoning: The Logic of Terms*. Ashgate, 2000.

[10] WANG, P. Return to term logic.

Many-Valued Oppositions

Yessica Espinoza Ramos
Benemérita Universidad Autónoma de Puebla

1 Introduction

The classical square of opposition, rooted in Aristotelian logic, has long served as a foundational framework for understanding logical relations between categorical propositions. Its four central relations—contradiction, contrariety, subcontrariety, and subalternation—rely on the bivalence and monotonicity of classical logic. However, the emergence of non-classical logics—such as three-valued logics (K3, Ł3), paraconsistent logics (LP, RM3), and logics of partial information (FDE)—has challenged the universality of this traditional schema. These systems introduce either a third truth value, the possibility of truth-value gaps or gluts, or both, thereby altering the semantic behavior of basic logical connectives.

This work investigates whether the propositional square of opposition can be preserved in these alternative systems. Through a systematic comparison of five non-classical logics—K3, Ł3, LP, RM3, and FDE—we show that only LP and RM3 maintain the classical square in full. The other systems either preserve some relations partially or invalidate the square altogether.

To address these incompatibilities, we propose a generalization of the square of opposition, reformulating each oppositional relation using a semantic framework based on positive and negative satisfaction ($\models^+$ and $\models^-$). This formalism enables a coherent extension of the classical oppositional structure to logics that tolerate indeterminacy, inconsistency, or both.

The structure of the paper is as follows. We begin by presenting the foundation of our study: the classical square of opposition within classical propositional logic. We then proceed to test the behavior of the classical square within each of the five systems under review, accompanied by truth tables. We then introduce our generalized definitions of the aristotelian relations and demonstrate their validity across non-classical logics. Finally, we present an enriched oppositional diagram that subsumes the classical case as a special instance while remaining applicable to a broader range of logical environments.

2 Classical Propositional Logic and Classical Square of Opposition

The language of Classical Propositional Logic (CPL) is composed of an indefinite number of variables, which can be uppercase or lowercase letters such as $A, B, C, \ldots, p, q, r, s, \ldots$. The variables represent sentences written in natural language; for example, the letter A can represent the sentence "snow is white", and the letter B can represent the sentence "the sky is blue". CPL also includes constants, which are the connectives: disjunction ($\vee$), conjunction ($\wedge$), negation ($\neg$), biconditional ($\equiv$), and conditional ($\rightarrow$). The semantics of CPL consists of two truth values, true (T) and false (F); therefore, it is bivalent.

In CPL, there are two types of propositions: atomic, which represent a sentence by means of a single variable —for example: "snow is white" (A)—, and molecular, which are propositions joined by one or more connectives —for example: "snow is white and the sky is blue" ($A \wedge B$), where B represents the sentence "the sky is blue" and the connective $\wedge$ corresponds, in natural language, to the word "and". The rule for producing atomic or molecular formulas is as follows: If A and B are formulas, then so are $\neg A$, $(A \wedge B)$, $(A \vee B)$, $(A \rightarrow B)$, and $(A \equiv B)$.

According to Malinowski, "Truth-tables determine the role and the meaning of the propositional connectives assigning to them functions whose arguments and values are 0 (zero) and 1 (one), denoting respectively falsity and truth."[1]

Truth tables assign truth values to each formula, an assignment is a function f that assigns a truth value to each propositional letter.

In this logical system, a proposition must have exactly one truth value, it is either true or false, but it cannot lack a truth value, nor can it possess both values simultaneously. The following table shows the connectives and their truth tables.

$\neg A$		$A \wedge B$			$A \vee B$			$A \rightarrow B$			$A \equiv B$		
		1	1	1	1	1	1	1	1	1	1	1	1
1	0	1	0	0	1	0	1	1	0	0	1	0	0
0	1	0	1	0	0	1	1	0	1	1	0	1	0
		0	0	0	0	0	0	0	0	1	0	0	1

Table 1: Truth Tables for Classical Connectives

---

[1][13, pp. 5–6].

In classical logic, semantics are based on the assignment of one of two truth values — 1 (true) or 0 (false) — to each atomic proposition. These values extend to molecular formulas according to the standard truth tables for the logical connectives. A valuation is a function f that assigns a truth value to each propositional letter.

A formula is said to be true under a valuation if, when truth values are assigned according to the rules of the truth tables, the formula evaluates to 1. A formula is a tautology if it is true under every possible valuation.

Given a set of formulas Γ and a formula φ, we say that φ is a logical consequence of Γ, written $\Gamma \vDash \varphi$, if and only if every valuation that makes all formulas in Γ true also makes φ true.

Tautologies play a central role in classical propositional logic. They are formulas that are true under every possible valuation, regardless of the truth values assigned to their components. As G. Malinowski puts it, "The tautologies of CPL are the laws of classical logic expressed in propositional language" [13, p. 6]. In this sense, tautologies serve as the backbone of classical reasoning: they capture necessary logical truths that hold universally, making them a key tool in the formal analysis of valid argument forms.

Name	Tautology
Double Negation	$\neg\neg A \equiv A$
Law of the Excluded Middle	$A \vee \neg A$
Law of Non-Contradiction	$\neg(A \wedge \neg A)$
Modus Ponens	$((A \rightarrow B) \wedge A) \rightarrow B$
Modus Tollens	$((A \rightarrow B) \wedge \neg B) \rightarrow \neg A$

Table 2: Examples of Classical Tautologies

In summary, classical propositional logic provides a well-defined framework for evaluating the truth of formulas through valuations and truth tables. Its tautologies express the fundamental laws of logical reasoning and serve as the basis for valid inference. With this semantic foundation in place, we can now turn to an analysis of the Classical Square of Opposition, which explores the relations of logical opposition between different types of propositions.

Although it is uncertain whether Aristotle himself conceived of a diagrammatic representation, a now-canonical diagram—known as the Square of Opposition—illustrates the structure of his theory of oppositions. According to this framework, the negative propositions, whether universal or particular, are derived from their corresponding affirmative counterparts.

Later, during the Middle Ages, a more formal representation of Aristotle's oppositional relations emerged. Philosophers such as Boethius[2] and Apuleius[3] developed what came to be known as the Square of Opposition, incorporating not only the relations Aristotle explicitly mentioned, but also additional oppositions. One of these is subcontrariety, a relation they inferred from Aristotle's remark that "in contrast with [the contraries], others may be opposed in relation to the same thing" [1, 17b 23—24]. They also identified an asymmetrical relation between the universal affirmative and the particular affirmative, and similarly between the universal negative and the particular negative. This relation came to be known as subalternation.

Some of Aristotle's successors—namely Apuleius and Boethius—proposed, centuries later, a compact geometric device to visually convey the fundamental logical principles articulated by Aristotle (such as the laws of non-contradiction and excluded middle). This diagram came to be known as the "Square of Opposition" (also referred to as the Logical Square, Apuleius' Square, or Aristotle's Square)[4]. Thus, a complete version of the square can be represented as in the Figure 1 below.

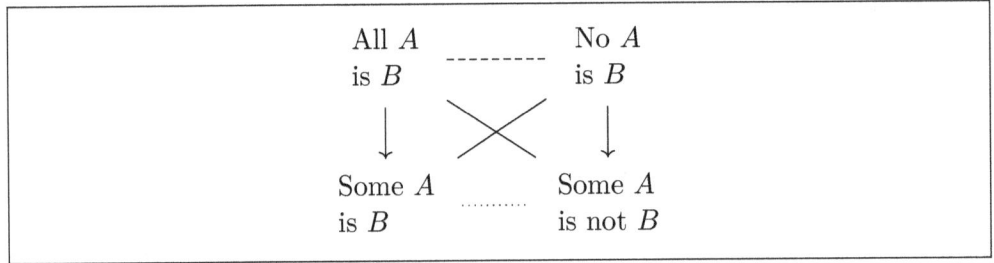

Figure 1: Classical Square of Opposition for Quantified Statements

Later on, from the figure of the square itself, a way of naming each of its corners emerged: to the two universal propositions, the affirmative and the negative, correspond the labels A and E respectively; and to the particular affirmative and particular negative, the labels I and O. As Redmond explains, "the scholastics used the following symbols to represent propositional forms ('a' and 'i' are the vowels of *affirmo* in Latin; 'e' and 'o' those of *nego*...)" [19, p. 89].

Manuel Correia, in *La lógica de Aristóteles*, explains why these particular vowels were chosen for the square: "The square is the work of the ancient

---
[2]Cf. [14].
[3]Cf. [11].
[4]Cf. [15, p. 78].

commentators, while the assignment of the letters A, E, I, and O is due to the medieval commentators. Boethius (*in Int.* 2, p. 152) and Ammonius (*in Int.*, p. 93) reproduce these squares in their commentaries" [8, p. 74].

Regarding the use of colors to distinguish the oppositional relations within the square, these were proposed by Jean-Yves Béziau: "contradiction is expressed in red, contrariety in blue, subcontrariety in green, and subalternation in black" [3, p. 220].

In recent years, numerous studies have been devoted to the classical square of opposition and its extensions, including developments such as logical hexagons and even three-dimensional figures. Entire books have been published where the main topic is the exploration of new insights into this theory[5].

From the development of classical propositional logic, various authors have sought to extend and reinterpret the traditional Aristotelian structures of opposition. One of the most notable cases is that of the Polish logician József Maria Bocheński (1902–1995), who, in his work *A Precis of Mathematical Logic*, specifically in the section titled *The Logic of Sentences*, provides a detailed explanation of truth tables and how they are constructed. Based on this formal perspective, Bocheński proposes a version of the square of opposition using logical connectives and Polish notation, thus transferring the traditional relations of contradiction, contrariety, subcontrariety, and subalternation into the realm of propositional logic. Below (Fig. 2) is our the propositional square of opposition formulated by Bocheński[6] [4, p. 14].

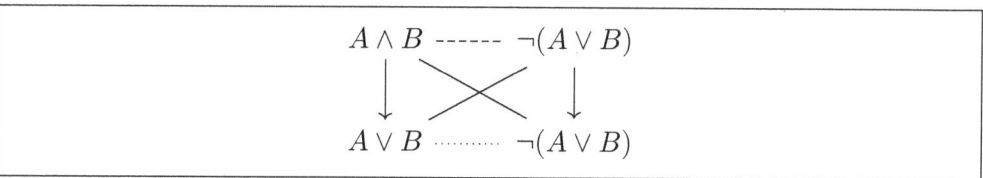

Figure 2: Classical Square of Opposition in CPL

We now present our own proposal for defining the classical relations of opposition, but rearticulated using logical tautologies. Each opposition is first introduced informally and then expressed formally as a tautology in the language of classical propositional logic. This approach allows us to link the semantic behavior of propositions to the structural relations of the traditional square of

---

[5]See [7, 5, 6, 16, 9].

[6]In the original version, Bochenski uses Polish (prefix) notation. However, in contemporary contexts, this notation has largely fallen out of use. For this reason, we have chosen to present the square in its modern version.

opposition.

- **Contradiction** Two propositions are contradictory if they cannot both be true and cannot both be false. *Formal expression:* $\models_{\mathsf{CPL}} \neg(\varphi \equiv \psi)$

$\neg$	$(\varphi$	$\equiv$	$\psi)$
0	1	1	1
1	1	0	0
1	0	0	1
0	0	1	0

Explanation: The table highlights in red the rows where the formula is true, thus considering only the cases that verify the condition—namely, where φ and ψ have different truth-values: if one is true, the other is false, and vice versa.

- **Contrariety** Two propositions are contraries if they cannot both be true, but they can both be false. *Formal expression:* $\models_{\mathsf{CPL}} \neg(\varphi \wedge \psi)$ and $\not\models_{\mathsf{CPL}} \varphi \vee \psi$

$\neg$	$(\varphi$	$\wedge$	$\psi)$	φ	$\vee$	ψ
0	1	1	1	1	1	1
1	1	0	0	1	1	0
1	0	0	1	0	1	1
1	0	0	0	0	0	0

Explanation: In this case, only the left-hand condition is satisfied according to the definition. Therefore, the rows in blue show the cases where the left-hand formula is true, that is, when one is true and the other is false. This condition makes the right-hand formula false in one case, which is why the disjunction does not hold. This aligns with the fact that contraries cannot both be true at the same time.

- **Subcontrariety** Two propositions are subcontraries if they cannot both be false, but they can both be true. *Formal expression:* $\models_{\mathsf{CPL}} \varphi \vee \psi$ and $\not\models_{\mathsf{CPL}} \neg(\varphi \wedge \psi)$

$\neg$	$(\varphi$	$\wedge$	$\psi)$	φ	$\vee$	ψ
0	1	1	1	1	1	1
1	1	0	0	1	1	0
1	0	0	1	0	1	1
1	0	0	0	0	0	0

Explanation: In this case, only the right-hand condition is satisfied, meaning that subcontraries can satisfy a disjunction. Therefore, the rows in green highlight these cases. As a result, in the left-hand formula, there is one case where the formula is false, when both are true. Both can be true at the same time, but not false at the same time.

- **Subalternation** One proposition implies another, but not vice versa. *Formal expression:* $\models_{CPL} \varphi \to \psi$ and $\not\models_{CPL} \psi \to \varphi$

φ	$\to$	ψ	ψ	$\to$	φ
1	1	1	1	1	1
1	0	0	0	1	1
0	1	1	1	0	0
0	1	0	0	1	0

Explanation: In this case, the tables show how subalternation is satisfied. On the left-hand side, the condition is met in the rows marked in gray, with the only case omitted being when the antecedent is true and the consequent is false. On the right-hand side, the formula fails in one of these gray-marked cases, establishing the fact that subalternation is a one-way conditional.

To sum up, we have revisited the classical square of opposition through the lens of propositional logic, using truth tables and formal definitions to clarify the traditional logical relations. This sets the stage for the next section, where we will explore how these relations behave in five different logical systems.

3 Five Systems and the Classical Square

In this section, we provide a general overview of five non-classical logical systems: Łukasiewicz's three-valued logic (Ł3), Kleene's three-valued logic (K3), the relevance logic RM3, the Logic of Paradox (LP), and First-Degree Entailment (FDE). Each of these systems was developed with different motivations, such as addressing vagueness, relevance, or contradictions. While we briefly mention these motivations, our primary focus will be on the structure of their respective truth tables. This groundwork will allow us to later analyze how the classical square of opposition behaves within each of these systems.

The motivation behind Jan Łukasiewicz's development of the first many-valued logical system was his rejection of determinism. As he defines it, determinism is the belief that if a proposition A is true at a moment t, then it has always been true that A is true at t. This implies that present events were already true before occurring, and future events must necessarily happen. Łukasiewicz identifies two main arguments for determinism: the principle of the excluded middle (which states that of two contradictory propositions, one must be true), and the principle of causality (which claims that every event must have a cause). He argues against determinism by noting that a future event cannot be known with certainty in the present, as intervening circumstances may alter the outcome. For him, the roots of this problem lie in Aristotle's *De Interpretatione*, Book IX [1], which discusses the famous example: "either there will be a sea battle tomorrow or there will not be a sea battle tomorrow." The principle of the excluded middle would suggest that one of these must necessarily happen. Łukasiewicz challenges this view by introducing a third truth-value to represent possibility: a statement about a future contingent event is neither true nor false, but rather possible. Thus, between truth and falsity, there exists a third value—*the possible*.

The three-valued logic system Ł3 shares some common features with **CPL**. For instance, it uses the same propositional variables (A, B, C, ...) and the same logical constants ($\neg$, $\wedge$, $\vee$, $\rightarrow$, $\equiv$). However, Ł3 introduces a third truth-value, making it a genuinely three-valued system. In addition to the classical values of true (1) and false (0), it includes the value of *possible* ($\frac{1}{2}$). As Łukasiewicz explains: "Three-valued logic is a system of non-Aristotelian logic, since it operates on the basis that, apart from true and false propositions, there are also propositions which are neither true nor false, and therefore a third value exists" [12].

A	B	$\neg A$	$A \wedge B$	$A \vee B$	$A \rightarrow B$	$A \equiv B$
1	1	0	1	1	1	1
1	$\frac{1}{2}$	0	$\frac{1}{2}$	1	$\frac{1}{2}$	$\frac{1}{2}$
1	0	0	0	1	0	0
$\frac{1}{2}$	1	$\frac{1}{2}$	$\frac{1}{2}$	1	1	$\frac{1}{2}$
$\frac{1}{2}$	$\frac{1}{2}$	$\frac{1}{2}$	$\frac{1}{2}$	$\frac{1}{2}$	1	1
$\frac{1}{2}$	0	$\frac{1}{2}$	0	$\frac{1}{2}$	$\frac{1}{2}$	$\frac{1}{2}$
0	1	1	0	1	1	0
0	$\frac{1}{2}$	1	0	$\frac{1}{2}$	1	$\frac{1}{2}$
0	0	1	0	0	1	1

Figure 3: Ł3 Truth-tables

A	B	$\neg A$	$A \wedge B$	$A \vee B$	$A \to B$
1	1	0	1	1	1
1	u	0	u	1	u
1	0	0	0	1	0
u	1	u	u	1	1
u	u	u	u	u	u
u	0	u	0	u	u
0	1	1	0	1	1
0	u	1	0	u	1
0	0	1	0	0	1

Figure 4: K3 truth-tables

In the Ł3 system, as in **CPL**, the designated value is 1. This means that a formula is considered a tautology only if it takes the value 1 in all possible cases of its truth table. Despite introducing an additional truth value to capture the notion of possibility, the criterion for logical validity remains conservative: only those propositions that are fully true under all interpretations are accepted as logically valid. In this way, Ł3 preserves continuity with **CPL** in terms of its notion of logical truth, while extending the semantic framework to account for indeterminacy. In Figure 3 truth tables are expressed.

Our next system **K3**, introduced in Stephen Cole Kleene's (1909–1994) work *Introduction to Metamathematics* [10]. In section §64, titled "The 3-valued logic," Kleene presents a new interpretation of the logical connectives, noting that some of them will be "undefined[7]." He writes: "it will be convenient to use three tables, with three truth-values t (true), f (false), and u (undefined), to describe the meanings which the connectives are now to have." [10, p. 332]

In this system, the third truth-value is u, standing for "undefined." Kleene clarifies that u is not the set of t and f, but rather the absence of information: "u merely means the absence of information whether $Q(x)$ is t or f." [10, p. 333] Later in the same section, he distinguishes between two types of truth tables: weak and strong. A weak table assigns the value u to a connective whenever u appears in any of the inputs; in contrast, the strong table follows the same reasoning with infimum and supremum as previously discussed. Figure 3 shows the truth tables for the *strong* version.

---

[7] While Kleene uses the symbols t and f for true and false, respectively, we will continue to use 1 and 0, as in the case of Ł3. Likewise, we will remain consistent with the third (or fourth) truth-value used in each system.

As in the Łukasiewicz system, the designated value in Kleene's three-valued logic is also 1, meaning that a formula is a tautology if and only if it evaluates to 1 in every row of the truth table. The only significant departure from Ł3 lies in the treatment of the conditional. In K3, the conditional is defined differently, reflecting a more cautious approach to inference when information is incomplete. This subtle shift illustrates how the interpretation of truth values, particularly the third value u, can reshape the logical behavior of familiar connectives while preserving core logical intuitions. We now turn to two additional systems that share many structural and semantic similarities with both Ł3 and K3, yet offer distinct perspectives on the treatment of indeterminacy and logical consequence.

The paraconsistent logic **LP** (Logic of Paradox), developed by the philosopher and logician Graham Priest, arises as a response to the challenge posed by semantic paradoxes—especially the Liar Paradox. Priest begins his exploration with a critical reflection on the nature of paradoxes:

> A paradox is an argument with premises which appear to be true and steps which appear to be valid, which nevertheless ends in a conclusion which is false. A solution would tell us which premise is false or which step invalid; but moreover, it would give us an independent reason for believing the premise or the step to be wrong. If we have no reason for believing the premise or the step other than that it blocks the conclusions, then the "solution" is *ad hoc* and unilluminating. [18, p. 220]

This passage underscores Priest's concern with traditional solutions to paradoxes that rely on rejecting intuitively compelling premises or inference steps solely to avoid contradiction. According to Priest, such responses lack explanatory power if they fail to provide independent justification beyond the mere need to preserve consistency. LP emerges from the recognition that some contradictions—particularly those arising in semantic contexts—might be true without collapsing the entire logical system. Thus, LP challenges the classical principle of explosion, embracing the possibility of true contradictions (dialetheias) and proposing a formal framework in which they can be meaningfully accommodated.

Technically, **LP** is a three-valued logic, similar in formal structure to Łukasiewicz's and Kleene's systems, yet with a distinct semantic motivation. Graham Priest introduces **LP** as a paraconsistent logic, where contradictions are permitted without collapsing the system into triviality. The three truth values are: 1 (true), 0 (false), and p (paradoxical), the latter being used by Priest to denote a value that is both true and false. While other systems, like Kleene's K3, interpret the intermediate value as a lack of information, LP treats p as the presence of conflicting information—that is, a genuine contradiction.

The following tables in Figure 5 present the truth-functional behavior of the standard logical connectives in LP.

A	B	$\neg A$	$A \wedge B$	$A \vee B$	$A \to B$
1	1	0	1	1	1
1	p	0	p	1	p
1	0	0	0	1	0
p	1	p	p	1	1
p	p	p	p	p	p
p	0	p	0	p	0
0	1	1	0	1	1
0	p	1	0	p	1
0	0	1	0	0	1

Figure 5: LP truth-tables

What sets this system apart from the previous ones is that LP adopts two designated values: t (true only) and p (paradoxical). The value p is both true and false, representing a kind of truth that tolerates contradiction. Accordingly, a formula is a logical truth if its main connective yields only designated values—t or p—in every row of its truth table. This broader criterion for logical truth reflects the paraconsistent nature of LP, which allows for contradictions without trivialization.

Another three-valued logic system is RM3, developed by Alan Ross Anderson (1925–1973) and Nuel D. Belnap (1930–2024). Like the previously discussed systems, RM3 is composed of propositional variables $(A, B, C, \ldots)$ and logical connectives $(\neg, \wedge, \vee, \to)$. In RM3, truth values are represented numerically: -1 for false, $+1$ for true, and 0 as the third intermediate value. While this numerical scheme differs from the symbols used in Łukasiewicz's or Kleene's systems, the idea remains similar: the value 0 is introduced to capture cases where truth is indeterminate or incomplete. Notably, RM3 is embedded within relevance logic, a broader framework aimed at preserving the relevance between premises and conclusions, and its truth tables are designed accordingly.

In the RM3 system, the set of designated values is $\{+1, 0\}$, meaning that both truth $(+1)$ and intermediate (0) are accepted as logically valid outcomes.

This feature distinguishes RM3 from other systems such as Ł3 or K3, where only a single value (or a distinct pair) is designated. Below in Figure 6 are the truth tables for the main logical connectives in RM3, following the conventions established above.

A	B	$\neg A$	$A \wedge B$	$A \vee B$	$A \rightarrow B$
+1	+1	−1	+1	+1	+1
+1	0	−1	0	+1	0
+1	−1	−1	−1	+1	−1
0	+1	0	0	+1	+1
0	0	0	0	0	+1
0	−1	0	−1	0	−1
−1	+1	+1	−1	+1	+1
−1	0	+1	−1	0	+1
−1	−1	+1	−1	−1	+1

Figure 6: RM3 truth tables

Finally, the four-valued logical system known as **FDE** (First-Degree Entailment) was developed by the American logicians Nuel D. Belnap (1930–2024), Alan Ross Anderson (1925–1973), and philosopher J. Michael Dunn (1941–2021). This system originated from an interest in programming machines capable of detecting contradictions through a dialogue-based format of questions and answers. Belnap introduces the motivation in clear terms:

> These four possibilities are precisely the four values of many-valued logic that I offer as a practical guide to computer reasoning. Let us give them names. The values are:
> T (only true),
> F (only false),
> *None* (neither true nor false), and
> *Both* (both true and false simultaneously). [2, p. 43]

The semantic structure of **FDE** is based on a four-valued lattice that models possible informational states regarding the truth of propositions. These four values—T (only true), F (only false), B (true and false), and N (neither true nor false)—form a non-linear lattice in which T and F are logically independent, while B is their join (least upper bound) and N is their meet (greatest lower bound). This lattice reflects how much information each value carries:

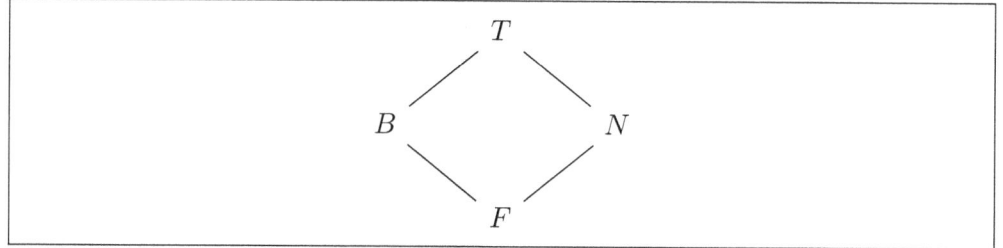

Figure 7: FDE four element lattice

N carries no information, T and F represent partial information, and B indicates informational overload, or contradiction. Logical operations in FDE are interpreted over this lattice structure, which accounts for the system's paraconsistent and paracomplete features. In Figure 8 we can see the truth tables of this logic[8].

In FDE, the set of designated values is T and B. This means that a formula is logically valid if, in every valuation, its main connective takes either the value T (true only) or B (both true and false). The inclusion of B among the designated values reflects FDE's paraconsistent approach: contradictions do not entail triviality. Instead, the system allows formulas to be both true and false, without compromising the structure of valid inference.

The systems examined—Ł3, K3, LP, RM3, and FDE—represent a spectrum of approaches to extending classical logic with additional truth-values. Ł3 and K3 are both three-valued systems that introduce an intermediate value: in Ł3, it reflects lack of information, whereas in K3 it denotes undefinedness; both retain a single designated value (1 or t). LP, in contrast, also uses a third value (denoted as p by Priest), but designates both the classical truth-value and the paradoxical one, aiming to model dialetheias without collapse into triviality. RM3 shares a similar motivation to LP but arises from relevance logic, employing numerical truth-values and designating both $+1$ and -1. Finally, FDE goes further, with four truth-values arranged in a lattice, capturing the interaction between truth and falsity explicitly; it designates both T (only true) and B (both true and false), embracing inconsistency and incompleteness. Despite their varied motivations—epistemic, semantic, or paraconsistent—these systems exhibit overlapping structural features, especially in their treatment of designated values and the behavior of negation and conditional. In what fol-

---

[8]It is worth noting that this system also admits an informational interpretation, which takes B as the top element and N as the bottom; however, this interpretation will not be considered in the present work due to space constraints.

Negation

A	$\neg A$
T	F
B	B
N	N
F	T

Conjunction

$\wedge$	T	B	N	F
T	T	B	N	F
B	B	B	F	F
N	N	F	N	F
F	F	F	F	F

Disjunction

$\vee$	T	B	N	F
T	T	T	T	T
B	T	B	T	B
N	T	T	N	N
F	T	B	N	F

Conditional

$\to$	T	B	N	F
T	T	B	N	F
B	T	B	T	B
N	T	T	N	N
F	T	T	T	T

Figure 8: **FDE** truth-tables

lows, we will explore how each of these logics reconfigures the classical square of opposition, offering new interpretations for contradiction, contrariety, and subalternation under non-classical semantics.

In the context of **FDE**, the classical square of opposition collapses. Since does not validate any tautologies, the traditional logical relations such as contradiction, contrariety, subcontrariety, and subalternation cannot be meaningfully expressed within the system. Consequently, the formulas that in classical logic denote relations of opposition (e.g., $\neg(\varphi \wedge \neg\psi)$ for contradiction) are invalid

> 1. $\not\models_{\mathsf{FDE,K3}} \neg((A \wedge B) \equiv \neg(A \wedge B))$
>
> 2. $\not\models_{\mathsf{FDE,K3}} \neg((A \vee B) \equiv \neg(A \vee B))$
>
> 3. $\not\models_{\mathsf{FDE,K3}} \neg((A \wedge B) \wedge \neg(A \vee B))$ and $\not\models_{\mathsf{FDE,K3}} (A \wedge B) \vee \neg(A \vee B)$
>
> 4. $\not\models_{\mathsf{FDE,K3}} \neg((A \vee B) \wedge \neg(A \wedge B))$ and $\not\models_{\mathsf{FDE,K3}} (A \vee B) \vee \neg(A \wedge B)$
>
> 5. $\not\models_{\mathsf{FDE,K3}} (A \wedge B) \to (A \vee B)$ and $\not\models_{\mathsf{FDE,K3}} (A \vee B) \to (A \wedge B)$
>
> 6. $\not\models_{\mathsf{FDE,K3}} \neg(A \vee B) \to \neg(A \wedge B)$ and $\not\models_{\mathsf{FDE,K3}} \neg(A \wedge B) \to (A \vee B)$

Figure 9: Aristotelian Relations in FDE and K3

in FDE. This reflects the paraconsistent and paracomplete nature of the logic: contradictions can be true, and excluded middles can fail, rendering the classical square inapplicable. In Figure 10, we can see the resulting square for FDE, which illustrates the absence of valid opposition relations.

The following list presents the expressions that are invalid in this system, that is, formulas that do not yield designated values in all possible cases.

> $(A \wedge B)$ $\neg(A \vee B)$
>
> $(A \vee B)$ $\neg(A \wedge B)$

Figure 10: Null Square in FDE and K3

The reader may verify that the truth table of the formulas expressing opposition relations is indeed not tautological. Although in most cases these formulas yield designated values, there is one specific case that refutes them: when both variables take the value N. In this situation, the formula fails to be designated, illustrating that the classical relations of opposition do not hold in FDE.

This same situation occurs in the logic K3. Since there are no tautologies in this system either, none of the traditional oppositional relations are valid. This is due to the fact that the third truth-value, u, systematically affects the validity of these logical relations, just as in FDE.

In contrast, the logic Ł3 does validate the subalternation relations of the classical square. Although K3 and Ł3 are essentially the same logic in terms of their truth values and general behavior, they differ in one key aspect: the definition of the conditional. This slight but significant difference has important consequences for the oppositional structure. In Ł3, thanks to the way conditional is defined, the subalternation relations—i.e., from universal to particular—are preserved. Therefore, formulas such as $(A \wedge B) \to (A \vee B)$ hold as valid, unlike in K3, where the same expression fails due to the weaker behavior of conditional.

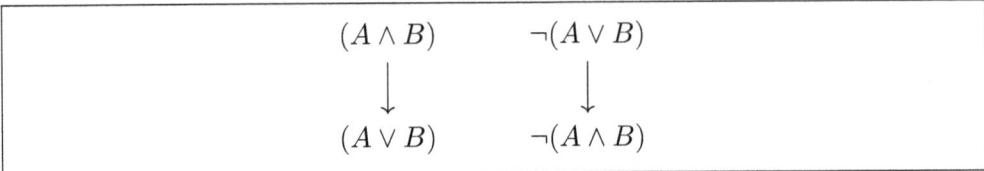

Figure 11: Degenerated SA Square in Ł3

In this case, the square is formed only by the two subalternation relations, as shown in Figure 11. The remaining oppositional relations—contradiction, contrariety, and subcontrariety—fail to hold in this logic. Below, we provide a list of valid and invalid relations, where the only distinction with respect to K3 lies in the validation of subalternation.

1. $\not\models_{Ł3} \neg((A \wedge B) \equiv \neg(A \wedge B))$
2. $\not\models_{Ł3} \neg((A \vee B) \equiv \neg(A \vee B))$
3. $\not\models_{Ł3} \neg((A \wedge B) \wedge \neg(A \vee B))$ and $\not\models_{Ł3} (A \wedge B) \vee \neg(A \vee B)$
4. $\not\models_{Ł3} \neg((A \vee B) \wedge \neg(A \wedge B))$ and $\not\models_{Ł3} (A \vee B) \vee \neg(A \wedge B)$
5. $\models_{Ł3} (A \wedge B) \to (A \vee B)$ and $\not\models_{Ł3} (A \vee B) \to (A \wedge B)$
6. $\models_{Ł3} \neg(A \vee B) \to \neg(A \wedge B)$ and $\not\models_{Ł3} \neg(A \wedge B) \to (A \vee B)$

Figure 12: Aristotelian Relations in Ł3

We now turn to the case of LP and RM3. In these logics, the square of opposition is exactly the same as in classical logic. Despite their paraconsistent character, all oppositional relations hold. The main reason for this lies in their sets of designated values: in both systems, the designated values include the classical truth value T and the intermediate value—p in Priest's system and 0 in RM3. In these cases, the intermediate value is interpreted as "both true and false," which stands in contrast to the interpretation in LP and Ł3, where

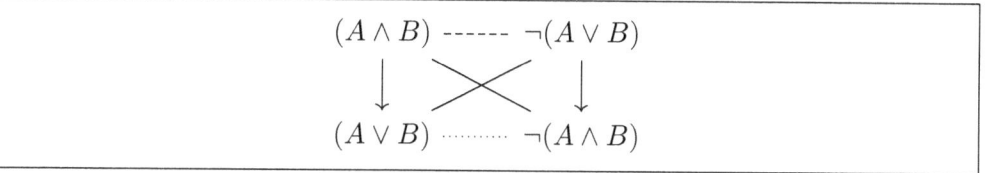

Figure 13: Classical Square in LP and RM3

the intermediate value is taken to mean "neither true nor false." The classical square holds precisely because the designated values preserve the necessary semantic structure for contradiction, contrariety, subcontrariety, and subalternation. The resulting square is shown in Figure 13, along with the corresponding list of valid relations. The reader may verify these facts by constructing the corresponding truth table.

1. $\models_{\text{LP,RM3}} \neg((A \wedge B) \equiv \neg(A \wedge B))$
2. $\models_{\text{LP,RM3}} \neg((A \vee B) \equiv \neg(A \vee B))$
3. $\models_{\text{LP,RM3}} \neg((A \wedge B) \wedge \neg(A \vee B))$ and $\not\models_{\text{LP,RM3}} (A \wedge B) \vee \neg(A \vee B)$
4. $\not\models_{\text{LP,RM3}} \neg((A \vee B) \wedge \neg(A \wedge B))$ and $\models_{\text{LP,RM3}} (A \vee B) \vee \neg(A \wedge B)$
5. $\models_{\text{LP,RM3}} (A \wedge B) \rightarrow (A \vee B)$ and $\not\models_{\text{LP,RM3}} (A \vee B) \rightarrow (A \wedge B)$
6. $\models_{\text{LP,RM3}} \neg(A \vee B) \rightarrow \neg(A \wedge B)$ and $\not\models_{\text{LP,RM3}} \neg(A \wedge B) \rightarrow (A \vee B)$

Figure 14: Aristotelian Relations in LP and RM3

After the preceding analysis, we can now draw a general conclusion: the classical square of opposition only holds in two of the five systems examined, namely LP and RM3. In the remaining three systems—Ł3, K3, and FDE—the classical relations of opposition fail to hold due to the semantic behavior of their intermediate or additional truth values. This naturally raises the question: does the propositional square of opposition retain its classical structure in non-classical logics? The answer is both yes and no. Yes, in LP and RM3, where the classical pattern remains valid; and no, in Ł3, K3, and FDE, where the structure collapses. What should we do, then? Rather than abandoning the classical relations of opposition, we propose to generalize them. In the next section, we introduce what we call the *many-valued square of opposition*, aimed at capturing the logical oppositions in systems where the classical model no longer applies.

4 The Many-Valued Square

In this section, we present a generalized version of the traditional opposition relations by making use of semantic notation adapted to many-valued logics. Specifically, we will employ the following conventions to express truth conditions for formulas[9]. We express the fact that a formula φ is verified as $\models^+ \varphi$; φ is not-verified as $\not\models^+ \varphi$; φ is falsified as $\models^- \varphi$; and φ is not-falsified as $\not\models^- \varphi$. This framework allows us to systematically redefine the classical square of opposition in a way that remains meaningful even in logics that lack tautologies, such as FDE, K3, and Ł3.

Let us begin with contradiction. We present our proposed generalization expressed using the notation introduced above. In addition, we include the truth table in FDE to verify that the proposed conditions are indeed satisfied.

- **Contradiction**: Two formulas φ and ψ are contradictory if and only if the negation of their conjunction cannot be falsified, and their disjunction cannot be falsified.

  Formal expression: $\not\models^-_{\mathsf{FDE}} \neg(\varphi \wedge \psi)$ and $\not\models^-_{\mathsf{FDE}} \varphi \vee \psi$

¬	(φ	∧	ψ)	φ	∨	ψ
F	T	T	T	T	T	T
B	T	B	B	T	T	B
N	T	N	N	T	T	N
V	T	F	F	T	T	F
B	B	B	T	B	T	T
B	B	B	B	B	B	B
T	B	F	N	B	T	N
T	B	F	F	B	B	F
N	N	N	T	N	T	T
T	N	F	B	N	T	B
N	N	N	N	N	N	N
T	N	F	F	N	N	F
T	F	F	F	F	T	T
T	F	F	B	F	B	B
T	F	F	N	F	N	N
T	F	F	F	F	F	F

---

[9]Cf, [17] and [20].

Explanation: In this case, the definition states that the formulas cannot be falsified. In both tables, the rows highlighted in red indicate the absence of the false value, which represents falsification. Although values such as N and B appear, they do not constitute a falsification of the formula. Therefore, both conditions are satisfied.

- **Contrariety**: Two formulas φ and ψ are contraries if and only if the negation of their conjunction cannot be falsified, and their disjunction can be neither falsified nor verified.

  Formal expression: $\not\models^-_{\mathsf{FDE}} \neg(\varphi \wedge \psi)$ and ($\not\models^-_{\mathsf{FDE}} \varphi \vee \psi$ and $\not\models^+_{\mathsf{FDE}} \varphi \vee \psi$)

¬	(φ	∧	ψ)	φ	∨	ψ
F	T	T	T	T	T	T
B	T	B	B	T	T	B
N	T	N	N	T	T	N
T	T	F	F	T	T	F
B	B	B	T	B	T	T
B	B	B	B	B	B	B
T	B	F	N	B	T	N
T	B	F	F	B	B	F
N	N	N	T	N	T	T
T	N	F	B	N	T	B
N	N	N	N	N	N	N
T	N	F	F	N	N	F
T	F	F	T	F	T	T
T	F	F	B	F	B	B
T	F	F	N	F	N	N
T	F	F	F	F	F	F

Explanation: In this case, only the formula on the left cannot be falsified, whereas the formula on the right can be neither verified (since there is a row with the value false) nor falsified (since there are rows with the value true).

- **Subcontrariety**: Two formulas φ and ψ are subcontraries if and only if the negation of their conjunction can be neither verified nor falsified, and their disjunction cannot be falsified.

  Formal expression: ($\not\models^+_{\mathsf{FDE}} \neg(\varphi \wedge \psi)$ and $\not\models^-_{\mathsf{FDE}} \neg(\varphi \wedge \psi)$) and $\not\models^-_{\mathsf{FDE}} \varphi \vee \psi$

	¬	(φ	∧	ψ)	φ	∨	ψ
	F	T	T	T	T	T	T
	B	T	B	B	T	T	B
	N	T	N	N	T	T	N
	T	T	F	F	T	T	F
	B	B	B	T	B	T	T
	B	B	B	B	B	B	B
	T	B	F	N	B	T	N
	T	B	F	F	B	B	F
	N	N	N	T	N	T	T
	T	N	F	B	N	T	B
	N	N	N	N	N	N	N
	T	N	F	F	N	N	F
	T	F	F	T	F	T	T
	T	F	F	B	F	B	B
	T	F	F	N	F	N	N
	T	F	F	F	F	F	F

Explanation: In this case, only the formula on the right cannot be falsified, as shown in the rows marked in green, while the formula on the left can be neither verified nor falsified, since there are rows with the values true, false, both (B), and neither (N).

- **Subalternation**: Two formulas φ and ψ are subalterns if and only if the conditional $\varphi \to \psi$ cannot be falsified, while the converse conditional $\psi \to \varphi$ can be neither verified nor falsified.

Formal expression: $\not\models^{-}_{FDE} \varphi \to \psi$ and ($\not\models^{-}_{FDE} \psi \to \varphi$ and $\not\models^{+}_{FDE} \psi \to \varphi$)

Many-Valued Oppositions

φ	$\rightarrow$	ψ	ψ	$\rightarrow$	φ
T	T	T	T	T	T
T	B	B	B	T	T
T	N	N	N	T	T
T	F	F	F	T	T
B	T	T	T	B	B
B	B	B	B	B	B
B	T	N	N	T	B
B	B	F	F	T	B
N	T	T	T	N	N
N	T	B	B	T	N
N	N	N	N	N	N
N	N	F	F	T	N
F	T	T	T	F	F
F	T	B	B	B	F
F	T	N	N	N	F
F	T	F	F	T	T

Explanation: In this case, since subalternation is characterized as a conditional, the rows in gray indicate where the conditional cannot be falsified, while on the right side, the formula with the conditional in the opposite direction can be neither verified nor falsified, as all four truth values appear.

Finally, to demonstrate the power of these definitions, we present the classical square in which the Aristotelian relations are satisfied in each system, and we also include the formulas according to our generalized definitions. The reader may verify that the conditions are fulfilled by constructing the corresponding truth tables. We will use S as an abbreviation for all the systems.

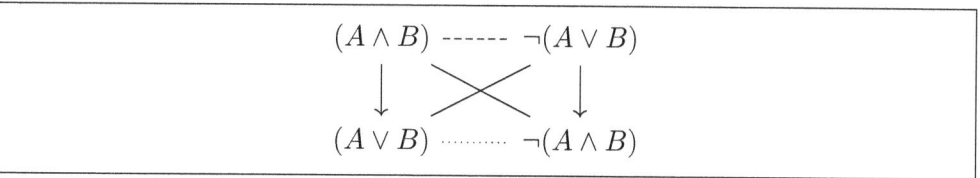

Figure 15: Classical Square in CPL, Ł3, K3, LP, RM3, and FDE

1. $\not\models_S^- \neg((A \wedge B) \wedge \neg(A \wedge B))$ and $\not\models_S^- (A \wedge B) \vee \neg(A \wedge B)$

2. $\not\models_S^- \neg((A \vee B) \wedge \neg(A \vee B))$ and $\not\models_S^- (A \vee B) \vee \neg(A \vee B)$

3. $\not\models_S^- \neg((A \wedge B) \wedge \neg(A \vee B))$ and ($\not\models_S^- (A \wedge B) \vee \neg(A \vee B)$ and $\not\models_S^+ (A \wedge B) \vee \neg(A \vee B)$)

4. ($\not\models_S^- \neg((A \vee B) \wedge \neg(A \wedge B))$ and $\not\models_S^+ \neg((A \vee B) \wedge \neg(A \wedge B))$) and $\not\models_S^- (A \vee B) \vee \neg(A \wedge B)$

5. $\not\models_S^- (A \wedge B) \to (A \vee B)$ and ($\not\models_S^- (A \vee B) \to (A \wedge B)$ and $\not\models_S^+ (A \vee B) \to (A \wedge B)$)

6. $\not\models_S^- \neg(A \vee B) \to \neg(A \wedge B)$ and ($\not\models_S^- \neg(A \wedge B) \to \neg(A \vee B)$ and $\not\models_S^+ \neg(A \wedge B) \to \neg(A \vee B)$)

Figure 16: Aristotelian Relations in CPL, Ł3, K3, LP, RM3, and FDE

5 Conclusion

Throughout this study, we have explored the behavior of the classical square of opposition across a range of non-classical logics, particularly those that admit a third truth value or allow for contradictions. We examined how logics such as Ł3, Ł3, LP, RM3, and Ł3 affect the traditional Aristotelian relations. Our analysis showed that only LP and RM3 preserve all classical oppositions, while in Ł3 only subalternation holds, and in K3 and FDE none of the classical square's relations are preserved in their traditional form.

To address this limitation, we introduced a generalized account of oppositional relations using the formal apparatus of positive and negative satisfaction ($\models^+$ and $\models^-$). This framework allows us to redefine contradiction, contrariety, subcontrariety, and subalternation in a way that remains valid across classical and non-classical systems alike. These generalized definitions are semantically grounded and compatible with truth-functional evaluations in multivalent and paraconsistent contexts.

This generalization not only preserves the intuitive core of Aristotelian opposition but also provides a robust tool for evaluating opposition in logical systems that diverge from classical assumptions. By presenting both the traditional and generalized squares, we have made explicit the precise conditions under which each relation holds and offered a path forward for future formal analyses of non-classical opposition.

References

[1] ARISTOTLE, . *Complete Works of Aristotle, Volume 1: the Revised Oxford Translation*. Princeton University Press, 1985.

[2] BELNAP, N. D. How a computer should think. *New essays on Belnap-Dunn logic* (2019), 35–53.

[3] BÉZIAU, J.-Y. New light on the square of oppositions and its nameless corner. *Logical Investigations 10* (01 2003), 218–232.

[4] BIRD, O., AND BOCHENSKI, J. *A Precis of Mathematical Logic*, vol. 1. Springer Science & Business Media, 2013.

[5] BÉZIAU, J.-Y. The power of the hexagon. *Logica Universalis 6*, 1 (2012), 1–43.

[6] BÉZIAU, J.-Y., AND GEROGIORGAKIS, S., Eds. *New Dimensions of the Square of Opposition*. Philosophia, 2017.

[7] BÉZIAU, J.-Y., AND JACQUETTE, D., Eds. *Around and Beyond the Square of Opposition*. Springer Science & Business Media, 2012.

[8] CORREIA, M. *La lógica de Aristóteles. Lecciones sobre el origen del pensamiento lógico en la antigüedad*. Santiago, Chile: Ediciones Universidad Católica de Chile, 2003.

[9] DEMEY, L. A hexagon of opposition for the theism/atheism debate. *Philosophia 47*, 2 (2019), 387–394.

[10] KLEENE, S. C. Introduction to metamathematics.

[11] LONDEY, D., AND JOHANSON, C. Apuleius and the square of opposition. *Phronesis 29* (1984), 165–173.

[12] ŁUKASIEWICZ, J. On three-valued logic. *The Polish Review* (1968), 43–44.

[13] MALINOWSKI, G. *Many-Valued Logics*. Oxford: Oxford University Press, 1993.

[14] MEISER, K. *Anicii Manlii Severini Boetii Commentarii in Librum Aristotelis Peri Ermineas 2*. Lipsiae: in aedibus B. G. Teubneri, 1880.

[15] MORETTI, A. *The geometry of logical opposition*. PhD thesis, Université de Neuchâtel, 2009.

[16] MORETTI, A. Why the logical hexagon? *Logica Universalis 6*, 1 (2012), 69–107.

[17] ODINTSOV, S. P., AND WANSING, H. Modal logics with belnapian truth values. *Journal of Applied Non-Classical Logics 20*, 3 (2010), 279–301.

[18] PRIEST, G. The logic of paradox. *Journal of Philosophical logic* (1979), 219–241.

[19] REDMOND, W. *Lógica simbólica para todos:(lógica elemental, modal, epistémica, deóntica, temporal y semántica de los mundos posibles)*. Universidad Veracruzana, 1999.

[20] SEDLÁR, I. Propositional dynamic logic with belnapian truth values. *arXiv preprint arXiv:1608.06084* (2016).

Logic-Sensitivity and Bitstring Semantics: From Squares to Hexagons of Opposition

Lorenz Demey
KU Leuven

StefFrijters
KU Leuven

1 Introduction

Over the past two decades, we have witnessed a renewed surge of interest in Aristotelian diagrams such as squares and hexagons of opposition [1, 2, 3, 4]. The research program of logical geometry studies such diagrams not only because of their historical and contemporary applications [7, 22, 23, 29], but also as objects of independent mathematical and philosophical interest [14, 20, 21, 31]. Two major research topics within this program concern bitstring semantics and logic-sensitivity of Aristotelian diagrams. By itself, bitstring semantics is well-understood and widely applied [13, 19, 34, 35, 36], and similar remarks apply to the topic of logic-sensitivity [11, 16, 19, 28]. However, the interplay between these two topics has turned out to be far more complicated than has long been thought. More concretely, bitstrings are not only combinatorial entities, but are typically also viewed as semantic representations of some formulas relative to some logical system. This reference to logical systems opens up a connection with the topic of logic-sensitivity: if the underlying logic of the formulas changes, how does this affect the bitstrings that represent those formulas? And vice versa, can every bitstring phenomenon that makes sense from a combinatorial perspective also be described from the semantic perspective, i.e., in terms of formulas and logics?

In a recent paper [17], we made an important first step toward addressing such questions. Initially adopting a purely combinatorial perspective on bitstrings, we described a very elegant account of deleting bit positions in squares of opposition. Then we switched to a semantic perspective on bitstrings, and asked whether this account could equally naturally be reformulated using formulas and logical systems. This was called 'Open Problem 1' in [17], and we discussed several potential solutions. However, none of these solutions was entirely successful. On the one hand, those that are based on natural, independently motivated formulas and logics, do not *completely* match the combinatorial account; on the other hand, those that do fully match the combinatorial account, are based on ad hoc logical systems, and thus feel rather *artificial*.

The present paper makes some new contributions toward this research line. We will argue that the aforementioned account can naturally be generalized from *squares* to *hexagons* of opposition, and explain how this helps to address a certain disadvantage from [17]. When dealing with hexagons of opposition, the interaction between logic-sensitivity and bitstring semantics continues to be very elegant at the purely combinatorial level, yet less satisfactory at the semantic level. We will also discuss how the move from squares to hexagons requires us to take into account an additional layer of complexity, viz., the existence of *Boolean subtypes* of Aristotelian families. After all, the account from [17] was entirely concerned with (classical and degenerate) squares of opposition, which do not have separate Boolean subtypes. By contrast, the generalized account to be developed here is concerned with hexagons of opposition, which do have multiple Boolean subtypes (e.g., *strong* vs. *weak* JSB hexagons)

The paper is organized as follows. Section 2 describes a combinatorially elegant account of deleting bit positions in squares of opposition, which was first presented in [17]. Next, Section 3 stays within the combinatorial perspective on bitstrings, and argues that this account should be generalized from squares to hexagons of opposition. Section 4 then switches to a more semantic perspective, and asks (under the label 'Open Problem 2') whether this story about hexagons of opposition can also be told in terms of formulas and logical systems, rather than just bitstrings. Section 5 exhibits concrete formulas and logical systems to show that this can indeed be done, but also argues that each solution to Open Problem 2 faces the same issue as those in [17]: completely matching the combinatorial account comes at the price of having to rely on ad hoc logical systems. Finally, Section 6 wraps things up by summarizing our main findings.

Before we get started with the paper, it also bears emphasizing that it has unfortunately not been possible to keep this paper entirely self-contained. Discussing and explaining all of the relevant notions and techniques (e.g., Aristotelian isomorphism, partition induced by a fragment, etc.) would easily have doubled the paper's length, while also increasing its overlap with [17] and other papers. The present paper constitutes a direct follow-up to [17], which does include detailed discussions of all the relevant prerequisites. Ideally, the reader is thus expected to be familiar with that paper; alternatively, they can fruitfully consult [19] or [12].

2 Deleting Bit Positions in Squares of Opposition

Motivated by considerations on modal logic (in particular, the logical systems K, KD and KF), Demey and Frijters [17] present the following problem, which is straightforward to describe from a purely combinatorial perspective on bitstrings. On the one hand, it is well-known in logical geometry that classical and degenerate squares of opposition can be represented by bitstrings of length 3 and 4, respectively. On the other hand, the easiest way to transform bitstrings of length 4 into bitstrings of length 3 involves simply deleting exactly one bit position. Since we start with four bit positions in total, there are thus four ways to transform a degenerate square into a classical square. This is fully described in Figure 1: part (a) of this figure shows a degenerate square, with bitstrings of length 4, while parts (b–e) show the four classical squares that result from deleting one bit position. More concretely, Figure 1 shows that every (deletion of a) bit position corresponds to a different direction of the subalternation arrows in the resulting classical square:

- Deleting the *first* bit position yields a classical square with the subalternations going *from left to right*; cf. Figure 1(b).

- Deleting the *second* bit position yields a classical square with the subalternations going *upwards*; cf. Figure 1(c).

- Deleting the *third* bit position yields a classical square with the subalternations going *from right to left*; cf. Figure 1(d).

- Deleting the *fourth* bit position yields a classical square with the subalternations going *downwards*; cf. Figure 1(e).

Although it may look deceivingly simple, this systematic correspondence between deleting bit positions and subalternation directions is of significant theoretical importance. However, we will not dwell on this importance at this point in the paper, because it was already discussed in detail in [17], and we will also return to it in Section 3, when we have broadened the scope from squares to hexagons.

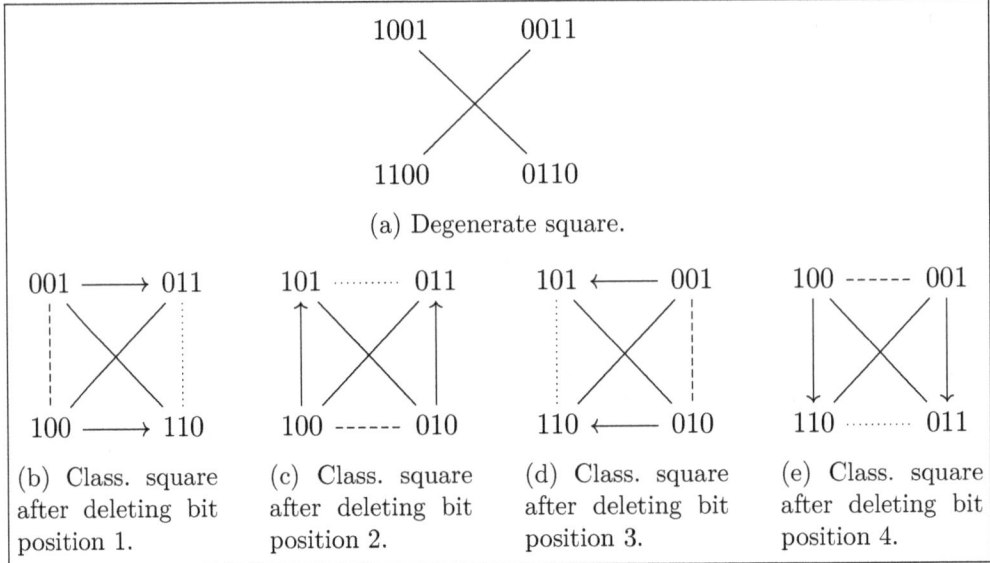

Figure 1: Five Aristotelian squares with purely combinatorial bitstrings.

The correspondence above was described from a purely *combinatorial* perspective: bitstrings and bit deletions were treated as purely combinatorial entities and operations, in complete isolation of any formulas or logics that they might be used to represent. However, in logical geometry, we typically take a thoroughly *semantic* perspective on bitstrings: bitstrings (and operations on them) are taken to represent the formulas of some fragment, relative to some ambient logical system. This immediately leads us to ask whether the correspondence between deleting bit positions and subalternation directions can not only be described from the combinatorial perspective, but also from the semantic perspective on bitstrings.

Demey and Frijters [17] show, first of all, that this combinatorial correspondence can indeed be expressed from the semantic perspective, and then go on to exhibit concrete formulas and logical systems to illustrate it. Once again, we will not discuss this in more detail right now, because after we have formulated a new combinatorial correspondence for hexagons of opposition (cf. Section 3), we will re-express that, too, from the semantic perspective (cf. Section 4), and then illustrate it by exhibiting concrete formulas and logical systems (cf. Section 5).

3 From Squares to Hexagons of Opposition

In the previous section, we have described (in purely combinatorial terms) a correspondence between deleting bit positions and the direction of the resulting subalternations. Our description has certain aspects of conventionality, which might seem to restrict its generality. For example, the correspondence only holds up to a permutation of the four bit positions (of the bitstrings that appear in the degenerate square in Figure 1(a)). Similarly, the correspondence is described in terms of (the directions of) the resulting subalternations, rather than any of the other resulting Aristotelian relations (i.e., contrarieties and subcontrarieties). These remarks are addressed in detail in [17], where it is argued that they are completely harmless. However, there is one aspect of conventionality which has hitherto not been addressed, and which is less innocent, in the sense that it does imply a genuine restriction on the generality of the correspondence between bit positions and subalternation directions.

The problem concerns the starting point we have chosen to describe the correspondence, viz., the degenerate square in Figure 1(a), which consists of the bitstrings 1001, 0110, 0011 and 1100. However, one can show that with bitstrings of length 4, one can construct *three* distinct degenerate squares:[1] next to that for {1001, 0110, 0011, 1100}, there is also a degenerate square for {1001, 0110, 1010, 0101} and one for {1010, 0101, 0011, 1100}. These three squares are shown in Figure 2, and since they are all degenerate, there is no principled reason for singling out one of them (in casu, the leftmost square from Figure 2) to act as the starting point for describing the correspondence between bit positions and subalternation directions (compare with Figure 1(a)). In other words: the story that was described in [17] could actually be told three times, each time taking a different degenerate square as our starting point.

In this paper, however, we will take a slightly different, more elegant route. Rather than looking at each of the three degenerate squares separately, we will combine them into a single hexagon, which simultaneously contains the six bitstrings 1001, 0110, 0011, 1100, 1010 and 0101, as shown in Figure 3(a). This diagram is a so-called *unconnectedness-12 (U12) hexagon*, since it contains 12 pairs of mutually unconnected bitstrings. The U12 hexagon is the obvious generalization of the degenerate square: while the latter consists of *two* pairs of contradictory formulas/bitstrings that do not stand in any other Aristotelian

---

[1]Some straightforward combinatorial considerations show that with bitstrings of length n, one can construct $\frac{1}{8}[4^n - 4 \cdot 3^n + 6 \cdot 2^n - 4]$ degenerate squares. Plugging in $n = 4$ into this formula yields the number of 3 degenerate squares that can be constructed with bitstrings of length 4.

relations, the former consists of *three* pairs of contradictory formulas/bitstrings that do not stand in any other Aristotelian relations. Furthermore, one can show that that there exists exactly *one* U12 hexagon that can be constructed with bitstrings of length 4;[2] hence, by taking this U12 hexagon as our new starting point, we are no longer forced to make some unprincipled or arbitrary choice.

If we systematically delete one bit position from all bitstrings in the U12 hexagon, we obtain a diagram with six bitstrings of length 3. It is well-known in logical geometry that the only kind of hexagon that can be constructed with bitstrings of length 3, is a *strong Jacoby-Sesmat-Blanché (JSB) hexagon*.[3] Since we start with four bit positions in total, there are thus four ways to transform the length-4 U12 hexagon into a strong JSB hexagon. This is fully described in Figure 3: part (a) of this figure shows the U12 hexagon with bitstrings of length 4, while parts (b–e) show the four strong JSB hexagons that result from deleting one bit position. These four resulting JSB hexagons can be differentiated in various ways; for example, one could refer to the directions of the resulting subalternations, or the locations of the resulting (sub)contrarieties, etc. This is another element of harmless conventionality, just like before. For the sake of concreteness, we will differentiate the four JSB hexagons in terms of the *location of their contrariety triangle*. From this perspective, Figure 3 thus shows that every (deletion of a) bit position corresponds to a different location of the contrariety triangle in the resulting JSB hexagon:

- Deleting the *first* bit position yields a strong JSB hexagon with the contrariety triangle in the *center* of the hexagon; cf. Figure 3(b).

- Deleting the *second* bit position yields a strong JSB hexagon with the contrariety triangle in the *upper left* part of the hexagon; cf. Figure 3(c).

- Deleting the *third* bit position yields a strong JSB hexagon with the contrariety triangle in the *lower* part the hexagon; cf. Figure 3(d).

---

[2]Again, some straightforward combinatorial considerations show that with bitstrings of length n, one can construct $\frac{1}{12}\left[2 \cdot 8^{n-1} - 3 \cdot 6^n + 6 \cdot 5^n - 2 \cdot 4^n - 6 \cdot 3^n + 8 \cdot 2^n - 4\right]$ U12 hexagons. Plugging in $n = 4$ into this formula yields the number of 1 U12 hexagon that can be constructed with bitstrings of length 4.

[3]This type of hexagon is named after Jacoby [24, 25], Sesmat [30] and Blanché [5, 6]. It is well-known that there are two Boolean subtypes of JSB hexagons: in a strong JSB hexagon, the disjunction of the three contraries is a tautology, whereas in a weak JSB hexagon, it is not [27]. In terms of bitstring semantics, strong JSB hexagons can be represented with bitstrings of length 3, while weak JSB hexagons require bitstrings of length 4 [19].

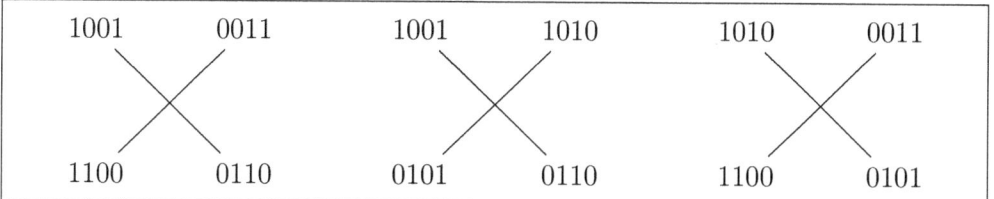

Figure 2: The three degenerate squares with bitstrings of length 4.

- Deleting the *fourth* bit position yields a strong JSB hexagon with the contrariety triangle in the *upper right* part of the hexagon; cf. Figure 3(e).

This new correspondence for *hexagons* is the natural generalization of the correspondence for *squares* that was first described in [17], and summarized in Section 2 of the present paper. After all, our starting point is now a U12 hexagon, which, as we have already seen above, naturally generalizes (and removes the arbitrariness of) the starting point of the previous correspondence, viz., a degenerate square. Furthermore, deleting bit positions now results in four strong JSB hexagons, which are the Boolean closures of the resulting diagrams of the previous correspondence, viz., four classical squares.

However, by going from squares to hexagons of opposition, we have also introduced a new layer of complexity. After all, it is well-known in logical geometry that squares of opposition do not have different Boolean subtypes [12, 19]: there exists only *one* type of degenerate squares (viz., those that can be represented with bitstrings of length 4), and there also exists only *one* type of classical squares (viz., those that can be represented with bitstrings of length 3). By contrast, most larger diagrams do have different Boolean subtypes [12, 13, 27]. In particular, there exist *five* types of U12 hexagons (viz., those that can be represented with bitstrings of length 8, 7, 6, 5 and 4, respectively), and there exist *two* types of JSB hexagons (viz., those that can be represented with bitstrings of length 4 and 3, also called 'weak' and 'strong', respectively). Hence, while the previous correspondence could be described purely in terms of Aristotelian families (degenerate square vs. classical square), for the new correspondence we need to refer to the Aristotelian families (U12 hexagon vs. JSB hexagon) but also to their Boolean subtypes (*length-4* U12 hexagon vs. *length-3/strong* JSB hexagon). For the combinatorial formulation in the present section, this is all fairly trivial, since the relevant bitstring lengths (i.e., the relevant Boolean subtypes) can immediately be observed in the diagrams in Figure 3. However, these considerations regarding Boolean subtypes do entail that we will have to be a bit more careful in Section 4, when we formulate the correspondence

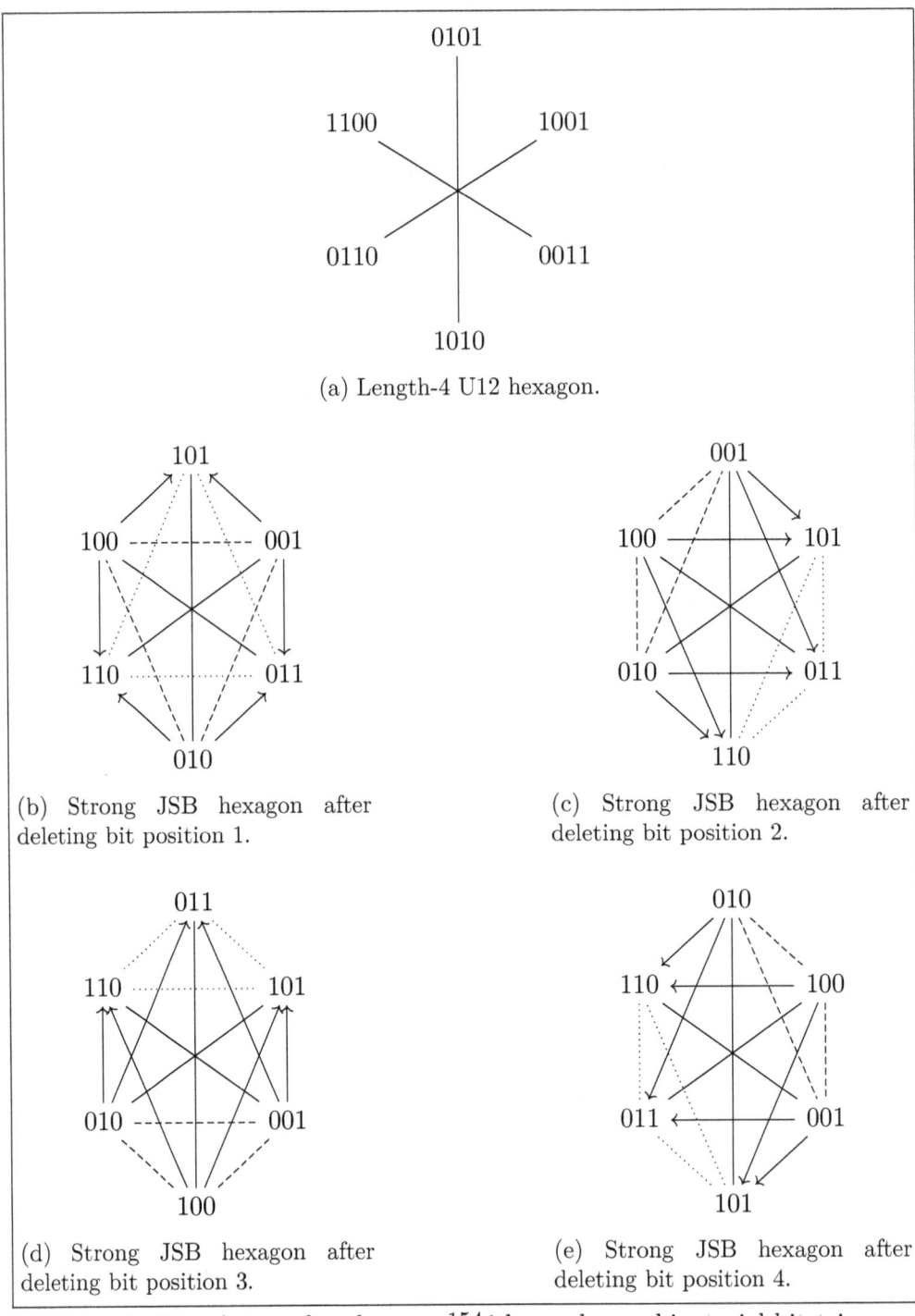

Figure 3: Five Aristotelian hexagons with purely combinatorial bitstrings.

from a semantic instead of a combinatorial perspective, i.e., in terms of logical systems instead of bitstrings.

To conclude this section, we emphasize that the systematic correspondence between bit positions and contrariety triangles brings together several major research lines in logical geometry, and is thus of significant theoretical importance. First of all, the correspondence is formulated in terms of Aristotelian families (e.g., U12 vs. JSB hexagon) and their Boolean subtypes (e.g., strong vs. weak JSB hexagons), and thus fits nicely within the ongoing research effort toward developing a systematic typology of Aristotelian diagrams [8, 13, 15, 26]. Secondly, the correspondence illustrates that bitstrings not only constitute a useful representation format [13, 35, 36], but are also theoretically fruitful for studying Aristotelian diagrams [12, 19, 34]. Thirdly, deleting bit positions are prime examples of 'infomorphisms', a new kind of morphism between Aristotelian diagrams that arises very naturally in the categorical perspective on logical geometry [9, 10, 37]. Finally, as was already explained in more detail in [17], the correspondence is also related to logical-geometrical correspondences in Aristotelian diagrams [18, 32] and the informativity ordering on logical relations [33].

4 A Semantic Reformulation of the Problem

In the previous section we have shown how the story about deleting bit positions from [17] can naturally be generalized from squares to hexagons of opposition, at least if we adopt a purely *combinatorial* perspective on bitstrings. However, as was already emphasized in [17], in logical geometry we typically view bitstrings as thoroughly *semantic* entities, i.e., as representations of some formulas relative to some logical system. In the present section, we therefore reformulate that story in terms of formulas and logical systems, rather than exclusively in terms of bitstrings. This semantic reformulation looks as follows.[4]

Open Problem 2. Find a language $\mathcal{L}$, a fragment $\mathcal{F} = \{\varphi, \psi, \chi, \neg\varphi, \neg\psi, \neg\chi\} \subseteq \mathcal{L}$, and five logical systems $\mathsf{S}_0, \mathsf{S}_1, \mathsf{S}_2, \mathsf{S}_3, \mathsf{S}_4$ for that same language $\mathcal{L}$, such that:

- The Aristotelian diagram for $(\mathcal{F}, \mathsf{S}_0)$ is a U12 hexagon such that the partition of S_0 that is induced by $\mathcal{F}$ is $\Pi_{\mathsf{S}_0}(\mathcal{F}) = \{\varphi \wedge \psi \wedge \chi, \varphi \wedge \neg\psi \wedge \neg\chi, \neg\varphi \wedge \psi \wedge \neg\chi, \neg\varphi \wedge \neg\psi \wedge \chi\}$; cf. Figure 4(a).

---

[4]The specific numbering has been chosen to facilitate the comparison between Open Problem 2 (pertaining to hexagons of opposition, present paper) with Open Problem 1 (pertaining to squares of opposition, [17]).

- The Aristotelian diagram for $(\mathcal{F}, \mathsf{S}_1)$ is a strong JSB hexagon, with pairwise contrarieties among φ, ψ, and χ; cf. Figure 4(b).

- The Aristotelian diagram for $(\mathcal{F}, \mathsf{S}_2)$ is a strong JSB hexagon, with pairwise contrarieties among φ, $\neg\psi$, and $\neg\chi$; cf. Figure 4(c).

- The Aristotelian diagram for $(\mathcal{F}, \mathsf{S}_3)$ is a strong JSB hexagon, with pairwise contrarieties among $\neg\varphi$, ψ, and $\neg\chi$; cf. Figure 4(d).

- The Aristotelian diagram for $(\mathcal{F}, \mathsf{S}_4)$ is a strong JSB hexagon, with pairwise contrarieties among $\neg\varphi$, $\neg\psi$, and χ; cf. Figure 4(e).

As was already explained in [17], ideally speaking we would also like the logical systems S_0, S_1, S_2, S_3 and S_4 to be *independently motivated* by their mathematical interest and/or philosophical applications. However, satisfying this additional criterion has turned out to be extremely challenging (even for squares of opposition, let alone hexagons). Hence, in this paper we will set this desideratum aside, and content ourselves with finding logical systems that meet all the formal requirements specified above, regardless of any further, independent motivations.

Note that the U12 hexagon for $(\mathcal{F}, \mathsf{S}_0)$ is not Aristotelian isomorphic to any of the JSB hexagons for $(\mathcal{F}, \mathsf{S}_n)$, for $1 \leq n \leq 4$. Using the terminology of [17], this yields four *blatant* cases of logic-sensitivity. Furthermore, note that for any $1 \leq m < n \leq 4$, the diagrams for $(\mathcal{F}, \mathsf{S}_m)$ and $(\mathcal{F}, \mathsf{S}_n)$ are both JSB hexagons (and are thus Aristotelian isomorphic to each other), but the identity function $id_\mathcal{F}$ is not an Aristotelian isomorphism between $(\mathcal{F}, \mathsf{S}_m)$ and $(\mathcal{F}, \mathsf{S}_n)$. In the terminology of [17], this thus yields six additional, more *subtle* cases of logic-sensitivity.[5]

Solving Open Problem 2 indeed amounts to 're-telling the story' from the previous section, but now from a semantic instead of a purely combinatorial perspective (i.e., in terms of logical systems instead of bitstrings). After all, for such a fragment $\mathcal{F} \subseteq \mathcal{L}$ and logical systems S_0, S_1, S_2, S_3, S_4, we have:

- $\Pi_{\mathsf{S}_0}(\mathcal{F}) = \{\varphi \wedge \psi \wedge \chi,\ \varphi \wedge \neg\psi \wedge \neg\chi,\ \neg\varphi \wedge \psi \wedge \neg\chi,\ \neg\varphi \wedge \neg\psi \wedge \chi\}$

---

[5]Our discussion on logic-sensitivity focuses exclusively on *Aristotelian* isomorphisms. For the sake of completeness, we also mention the analogous results for *Boolean* isomorphisms. The U12 hexagon for $(\mathcal{F}, \mathsf{S}_0)$ is not Boolean isomorphic to any of the JSB hexagons for $(\mathcal{F}, \mathsf{S}_n)$, for $1 \leq n \leq 4$. Furthermore, for any $1 \leq m < n \leq 4$, the JSB hexagons for $(\mathcal{F}, \mathsf{S}_m)$ and $(\mathcal{F}, \mathsf{S}_n)$ are both strong (and thus Boolean isomorphic to each other), but $id_\mathcal{F}$ is not a Boolean isomorphism between $(\mathcal{F}, \mathsf{S}_m)$ and $(\mathcal{F}, \mathsf{S}_n)$.

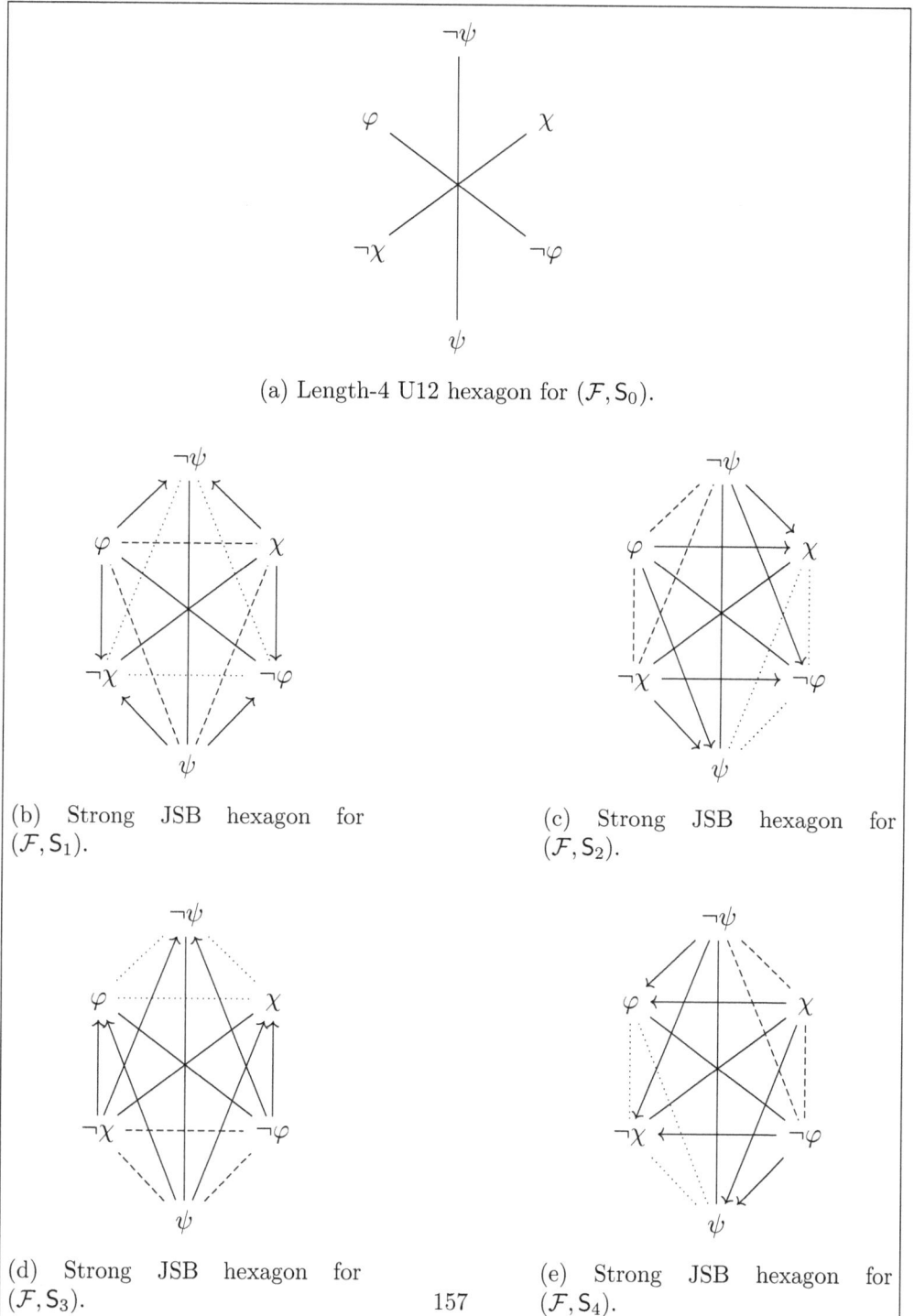

Figure 4: Five Aristotelian hexagons for Open Problem 2.

Since $\varphi \equiv_{S_0} (\varphi \wedge \psi \wedge \chi) \vee (\varphi \wedge \neg\psi \wedge \neg\chi)$, we have $\beta_{S_0}^{\mathcal{F}}(\varphi) = 1100$. Similarly, we find that $\beta_{S_0}^{\mathcal{F}}(\psi) = 1010$, $\beta_{S_0}^{\mathcal{F}}(\chi) = 1001$, $\beta_{S_0}^{\mathcal{F}}(\neg\varphi) = 0011$, $\beta_{S_0}^{\mathcal{F}}(\neg\psi) = 0101$ and $\beta_{S_0}^{\mathcal{F}}(\neg\chi) = 0110$. Compare the formulas in Figure 4(a) with the bitstrings in Figure 3(a).

- $\Pi_{S_1}(\mathcal{F}) = \{\varphi, \psi, \chi\}$

  If we compare $\Pi_{S_1}(\mathcal{F})$ with $\Pi_{S_0}(\mathcal{F})$, we find that the second, third and fourth anchor formulas have been simplified (viz., from $\varphi \wedge \neg\psi \wedge \neg\chi$ to φ, from $\neg\varphi \wedge \psi \wedge \neg\chi$ to ψ, and from $\neg\varphi \wedge \neg\psi \wedge \chi$ to χ, respectively), while its first anchor formula (viz., $\varphi \wedge \psi \wedge \chi$) is S_1-inconsistent, and is thus absent from $\Pi_{S_1}(\mathcal{F})$ altogether.

  An easy computation yields $\beta_{S_1}^{\mathcal{F}}(\varphi) = 100$, $\beta_{S_1}^{\mathcal{F}}(\psi) = 010$, $\beta_{S_1}^{\mathcal{F}}(\chi) = 001$, $\beta_{S_1}^{\mathcal{F}}(\neg\varphi) = 011$, $\beta_{S_1}^{\mathcal{F}}(\neg\psi) = 101$ and $\beta_{S_1}^{\mathcal{F}}(\neg\chi) = 110$. Compare Figure 4(b) with Figure 3(b).

- $\Pi_{S_2}(\mathcal{F}) = \{\varphi, \neg\chi, \neg\psi\}$

  If we compare $\Pi_{S_2}(\mathcal{F})$ with $\Pi_{S_0}(\mathcal{F})$, we find that the first, third and fourth anchor formulas have been simplified (viz., from $\varphi \wedge \psi \wedge \chi$ to φ, from $\neg\varphi \wedge \psi \wedge \neg\chi$ to $\neg\chi$, and from $\neg\varphi \wedge \neg\psi \wedge \chi$ to $\neg\psi$, respectively), while its second anchor formula (viz., $\varphi \wedge \neg\psi \wedge \neg\chi$) is S_2-inconsistent, and is thus absent from $\Pi_{S_1}(\mathcal{F})$ altogether.

  An easy computation yields $\beta_{S_2}^{\mathcal{F}}(\varphi) = 100$, $\beta_{S_2}^{\mathcal{F}}(\psi) = 110$, $\beta_{S_2}^{\mathcal{F}}(\chi) = 101$, $\beta_{S_2}^{\mathcal{F}}(\neg\varphi) = 011$, $\beta_{S_2}^{\mathcal{F}}(\neg\psi) = 001$ and $\beta_{S_2}^{\mathcal{F}}(\neg\chi) = 010$. Compare Figure 4(c) with Figure 3(c).

- $\Pi_{S_3}(\mathcal{F}) = \{\psi, \neg\chi, \neg\varphi\}$

  If we compare $\Pi_{S_3}(\mathcal{F})$ with $\Pi_{S_0}(\mathcal{F})$, we find that the first, second and fourth anchor formulas have been simplified (viz., from $\varphi \wedge \psi \wedge \chi$ to ψ, from $\varphi \wedge \neg\psi \wedge \neg\chi$ to $\neg\chi$, and from $\neg\varphi \wedge \neg\psi \wedge \chi$ to $\neg\varphi$, respectively), while its third anchor formula (viz., $\neg\varphi \wedge \psi \wedge \neg\chi$) is S_3-inconsistent, and is thus absent from $\Pi_{S_1}(\mathcal{F})$ altogether.

  An easy computation yields $\beta_{S_3}^{\mathcal{F}}(\varphi) = 110$, $\beta_{S_3}^{\mathcal{F}}(\psi) = 100$, $\beta_{S_3}^{\mathcal{F}}(\chi) = 101$, $\beta_{S_3}^{\mathcal{F}}(\neg\varphi) = 001$, $\beta_{S_3}^{\mathcal{F}}(\neg\psi) = 011$ and $\beta_{S_3}^{\mathcal{F}}(\neg\chi) = 010$. Compare Figure 4(d) with Figure 3(d).

- $\Pi_{S_4}(\mathcal{F}) = \{\chi, \neg\psi, \neg\varphi\}$

  If we compare $\Pi_{S_4}(\mathcal{F})$ with $\Pi_{S_0}(\mathcal{F})$, we find that the first, second and third anchor formulas have been simplified (viz., from $\varphi \wedge \psi \wedge \chi$ to χ, from $\varphi \wedge \neg\psi \wedge \neg\chi$ to $\neg\psi$, and from $\neg\varphi \wedge \psi \wedge \neg\chi$ to $\neg\varphi$, respectively), while its fourth anchor formula (viz., $\neg\varphi \wedge \neg\psi \wedge \chi$) is S_4-inconsistent, and is thus absent from $\Pi_{\mathsf{S}_1}(\mathcal{F})$ altogether.

An easy computation yields $\beta_{\mathsf{S}_4}^{\mathcal{F}}(\varphi) = 110$, $\beta_{\mathsf{S}_4}^{\mathcal{F}}(\psi) = 101$, $\beta_{\mathsf{S}_4}^{\mathcal{F}}(\chi) = 100$, $\beta_{\mathsf{S}_4}^{\mathcal{F}}(\neg\varphi) = 001$, $\beta_{\mathsf{S}_4}^{\mathcal{F}}(\neg\psi) = 010$ and $\beta_{\mathsf{S}_4}^{\mathcal{F}}(\neg\chi) = 011$. Compare Figure 4(e) with Figure 3(e).

These considerations clearly show that Open Problem 2 generalizes Open Problem 1 (as formulated in [17]) from squares to hexagons of oppositions. The formulations of both problems are highly similar to each other, with one crucial exception, viz., the importance of *Boolean subtypes*. Open Problem 1 from [17] concerns (degenerate and classical) squares of opposition, which do not have Boolean subtypes, and hence Open Problem 1 did not need to specify any further requirements in this regard (degenerate squares always require bitstring length 4; classical squares always require bitstring length 3). By contrast, Open Problem 2 from the present paper concerns (U12 and JSB) hexagons of opposition, which do have different Boolean subtypes, so Open Problem 2 does have to be more specific in this regard. More concretely, it does not suffice to specify the Aristotelian family of all the hexagons involved (viz., they have to be *U12* and *JSB* hexagons); we also have to specify their Boolean subtype (viz., they have to be *length-4* U12 and *strong/length-3* JSB hexagons). Furthermore, for the U12 hexagon for $(\mathcal{F}, \mathsf{S}_0)$, it does not even suffice to specify that it has to be representable by bitstrings of length 4 — i.e., $|\Pi_{\mathsf{S}_0}(\mathcal{F})| = 4$ —; we really have to specify what the four anchor formulas in $\Pi_{\mathsf{S}_0}(\mathcal{F})$ look like exactly. Since the diagram for $(\mathcal{F}, \mathsf{S}_0)$ is a U12 hexagon, the partition $\Pi_{\mathsf{S}_0}(\mathcal{F})$ can in principle contain up to eight anchor formulas, of the form $\pm\varphi \wedge \pm\psi \wedge \pm\chi$. Of these eight candidates, we require exactly the four that are mentioned in Open Problem 2 to be S_0-consistent (and thus to belong to $\Pi_{\mathsf{S}_0}(\mathcal{F})$), in order to guarantee the elegant correspondence between deleting bit positions and resulting contrariety triangles in the remainder of Open Problem 2. The fact that we have to specify the four anchor formulas in $\Pi_{\mathsf{S}_0}(\mathcal{F})$ can thus be viewed as a new harmless aspect of conventionality, just like those we encountered with Open Problem 1 in [17].

5 A Semantic Solution

The combinatorial stories that were told in [17] and in Section 3 of the present paper both concern (diagrams that can be encoded with) the same bitstring lengths, viz., bitstrings of length 4 to encode a degenerate square/U12 hexagon, and bitstrings of length 3 to encode classical squares/JSB hexagons. Consequently, it is easy to see that Open Problem 1 of [17] and Open Problem 2 of the present paper are co-solvable: every solution to one of these open problems can be transformed into a solution to the other one. Strategically speaking, this co-solvability is a double-edged sword. On the negative side, since every solution to Open Problem 2 automatically yields a solution to Open Problem 1 as well, the pessimism we voiced in [17] regarding finding a natural (not *ad hoc*) solution to Open Problem 1 immediately spreads to Open Problem 2 as well. However, on the positive side, since every solution to Open Problem 1 can straightforwardly be generalized to a solution to Open Problem 2, we will set aside our concerns regarding naturality, take the solution to Open Problem 1 that was presented in [17], and effectively generalize it to a solution to Open Problem 2.

First of all, we introduce the same language $\mathcal{L}_{\square\blacksquare}^{\circ\bullet}$ as in [17], which has, next to the usual Boolean connectives, four unary connectives $\circ$, $\bullet$, $\square$ and $\blacksquare$. It is thus defined by the following BNF:

$$\varphi ::= p \mid \varphi \wedge \varphi \mid \neg\varphi \mid \circ\varphi \mid \bullet\varphi \mid \square\varphi \mid \blacksquare\varphi$$

The other Boolean connectives and the duals of $\square$ and $\blacksquare$ are defined as usual; in particular, we have $\Diamond\varphi := \neg\square\neg\varphi$ and $\blacklozenge\varphi := \neg\blacksquare\neg\varphi$. We will work with the fragment $\mathcal{H}_{\square\blacksquare}^{\circ\bullet} := \{\varphi, \psi, \chi, \neg\varphi, \neg\psi, \neg\chi\}$, where

1. $\varphi := (\square\blacksquare p \wedge \Diamond\blacksquare\bullet p) \vee (\Diamond\blacksquare\bullet p \wedge \Diamond\circ\blacklozenge\bullet p)$,

2. $\psi := (\square\blacksquare p \wedge \Diamond\blacksquare\bullet p) \vee (\square\circ\blacklozenge p \wedge \Diamond\circ\blacklozenge\bullet p)$,

3. $\chi := (\square\blacksquare p \wedge \Diamond\blacksquare\bullet p) \vee (\square\blacksquare p \wedge \square\circ\blacklozenge p)$,

4. $\neg\varphi := (\square\circ\blacklozenge p \wedge \Diamond\circ\blacklozenge\bullet p) \vee (\square\blacksquare p \wedge \square\circ\blacklozenge p)$,

5. $\neg\psi := (\Diamond\blacksquare\bullet p \wedge \Diamond\circ\blacklozenge\bullet p) \vee (\square\blacksquare p \wedge \square\circ\blacklozenge p)$ and

6. $\neg\chi := (\Diamond\blacksquare\bullet p \wedge \Diamond\circ\blacklozenge\bullet p) \vee (\square\circ\blacklozenge p \wedge \Diamond\circ\blacklozenge\bullet p)$.

Finally, S_0^*, S_1^*, S_2^*, S_3^* and S_4^* are systems of bimodal logic, which are interpreted on Kripke models $\langle W, R^\square, R^\blacksquare, V\rangle$. These are precisely the same bimodal logics as in [17]. Concretely, this means that the modal operators $\square$ and $\blacksquare$ are interpreted in terms of the relations $R^\square$ and $R^\blacksquare$, respectively, and that the full semantics of these logics can be summarized by the following table:

LOGIC-SENSITIVITY AND BITSTRING SEMANTICS

	○	□	●	■
S_0^*	¬	K	id	id
S_1^*	id	id	¬	KD
S_2^*	¬	KF	id	id
S_3^*	id	id	¬	KF
S_4^*	¬	KD	id	id

The interpretation of this table is exactly the same as in [17]; for the sake of completeness, the explanatory comments of [17] are repeated here nearly verbatim. For all $\star \in \{\circ, \square, \bullet, \blacksquare\}$ and $S \in \{S_0^*, S_1^*, S_2^*, S_3^*, S_4^*\}$, the cell $\begin{array}{c|c} & \star \\ \hline S & id \end{array}$ in the table means that $\star$ is the identity connective in S, i.e., $\star\varphi \equiv_S \varphi$. In case $\star$ is a modal operator, $\square$ or $\blacksquare$, such an id-cell moreover entails that $\square\varphi \equiv_S \varphi \equiv_S \Diamond\varphi$ or $\blacksquare\varphi \equiv_S \varphi \equiv_S \blacklozenge\varphi$, respectively. Furthermore, for $\star \in \{\circ, \bullet\}$, the cell $\begin{array}{c|c} & \star \\ \hline S & \neg \end{array}$ means that $\star$ is a classical negation in S, i.e., $\star\varphi \equiv_S \neg\varphi$. Finally, for $\star \in \{\square, \blacksquare\}$, the cell $\begin{array}{c|c} & \star \\ \hline S & K \end{array}$ means that $\star$ is a K-type modal operator in S (and similarly for KD and KF).[6] For example, the second row of the table states that S_1^* is interpreted on Kripke models $\mathbb{M} = \langle W, R^\square, R^\blacksquare, V \rangle$ such that $R^\square$ is the identity relation on W, $R^\blacksquare$ is a serial relation on W, and the semantics for $\circ$ and $\bullet$ is: $\mathbb{M}, w \models \circ\varphi$ iff $\mathbb{M}, w \models \varphi$, while $\mathbb{M}, w \models \bullet\varphi$ iff $\mathbb{M}, w \not\models \varphi$.[7]

We can now prove the following:

- In S_0^*, the formulas of $\mathcal{H}_{\square\blacksquare}^{\circ\bullet}$ simplify as follows:

  - $\varphi = (\square\blacksquare p \wedge \Diamond\blacksquare \bullet p) \vee (\Diamond\blacksquare \bullet p \wedge \Diamond \circ \blacklozenge \bullet p)$
    $\equiv_{S_0^*} (\square p \wedge \Diamond p) \vee (\Diamond p \wedge \Diamond \neg p) \equiv_{S_0^*} \Diamond p$,
  - $\psi = (\square\blacksquare p \wedge \Diamond\blacksquare \bullet p) \vee (\square \circ \blacklozenge p \wedge \Diamond \circ \blacklozenge \bullet p)$
    $\equiv_{S_0^*} (\square p \wedge \Diamond p) \vee (\square \neg p \wedge \Diamond \neg p) \equiv_{S_0^*} \Diamond p \leftrightarrow \square p$,

---

[6]We say that $\square$ is a K- resp. a KD- resp. a KF-type modal operator iff $R^\square$ is a relation on W resp. a serial relation on W (i.e. $\forall w \in W : \exists v \in W : wR^\square v$) resp. a partially functional relation on W (i.e., $\forall w, v, v' \in W$: if $wR^\square v$ and $wR^\square v'$ then $v = v'$). Analogous remarks apply to $\blacksquare$ and $R^\blacksquare$.

[7]Note that in none of the logics considered here, $\circ$ and $\bullet$ are really necessary, in the sense that every formula containing $\circ$ or $\bullet$ can be rewritten as an equivalent formula that does not contain $\circ$ or $\bullet$. However, this equivalent formula will look different in the different logical systems. For example, $\circ p \equiv_S \neg p$ and $\bullet p \equiv_S p$ for $S \in \{S_0^*, S_2^*, S_4^*\}$, while $\circ p \equiv_S p$ and $\bullet p \equiv_S \neg p$ for $S \in \{S_1^*, S_3^*\}$.

- $\chi = (\Box\blacksquare p \land \Diamond\blacksquare \bullet p) \lor (\Box\blacksquare p \land \Box \circ \blacklozenge p)$
  $\equiv_{\mathsf{S}_0^*} (\Box p \land \Diamond p) \lor (\Box p \land \Box\neg p) \equiv_{\mathsf{S}_0^*} \Box p$,
- $\neg\varphi = (\Box \circ \blacklozenge p \land \Diamond \circ \blacklozenge \bullet p) \lor (\Box\blacksquare p \land \Box \circ \blacklozenge p)$
  $\equiv_{\mathsf{S}_0^*} (\Box\neg p \land \Diamond\neg p) \lor (\Box p \land \Box\neg p) \equiv_{\mathsf{S}_0^*} \Box\neg p$,
- $\neg\psi = (\Diamond\blacksquare \bullet p \land \Diamond \circ \blacklozenge \bullet p) \lor (\Box\blacksquare p \land \Box \circ \blacklozenge p)$
  $\equiv_{\mathsf{S}_0^*} (\Diamond p \land \Diamond\neg p) \lor (\Box p \land \Box\neg p) \equiv_{\mathsf{S}_0^*} \Diamond p \leftrightarrow \Diamond\neg p$,
- $\neg\chi = (\Diamond\blacksquare \bullet p \land \Diamond \circ \blacklozenge \bullet p) \lor (\Box \circ \blacklozenge p \land \Diamond \circ \blacklozenge \bullet p)$
  $\equiv_{\mathsf{S}_0^*} (\Diamond p \land \Diamond\neg p) \lor (\Box\neg p \land \Diamond\neg p) \equiv_{\mathsf{S}_0^*} \Diamond\neg p$.

Since $\Box$ is a K-modality in S_0^*, the Aristotelian diagram for $(\mathcal{H}_{\Box\blacksquare}^{\circ\bullet}, \mathsf{S}_0^*)$ is a U12 hexagon; cf. Figure 5(a). The partition $\Pi_{\mathsf{S}_0^*}(\mathcal{H}_{\Box\blacksquare}^{\circ\bullet})$ contains the following anchor formulas:

- $\Box\blacksquare p \land \Diamond\blacksquare \bullet p \equiv_{\mathsf{S}_0^*} \Box p \land \Diamond p \equiv_{\mathsf{S}_0^*} \varphi \land \psi \land \chi$,
- $\Diamond\blacksquare \bullet p \land \Diamond \circ \blacklozenge \bullet p \equiv_{\mathsf{S}_0^*} \Diamond p \land \Diamond\neg p \equiv_{\mathsf{S}_0^*} \varphi \land \neg\psi \land \neg\chi$,
- $\Box \circ \blacklozenge p \land \Diamond \circ \blacklozenge \bullet p \equiv_{\mathsf{S}_0^*} \Box\neg p \land \Diamond\neg p \equiv_{\mathsf{S}_0^*} \neg\varphi \land \psi \land \neg\chi$,
- $\Box\blacksquare p \land \Box \circ \blacklozenge p \equiv_{\mathsf{S}_0^*} \Box p \land \Box\neg p \equiv_{\mathsf{S}_0^*} \neg\varphi \land \neg\psi \land \chi$.

This shows that the diagram for $(\mathcal{H}_{\Box\blacksquare}^{\circ\bullet}, \mathsf{S}_0^*)$ is a length-4 U12 hexagon.

- In S_1^*, the formulas of $\mathcal{H}_{\Box\blacksquare}^{\circ\bullet}$ simplify as follows:

  - $\varphi = (\Box\blacksquare p \land \Diamond\blacksquare \bullet p) \lor (\Diamond\blacksquare \bullet p \land \Diamond \circ \blacklozenge \bullet p)$
    $\equiv_{\mathsf{S}_1^*} (\blacksquare p \land \blacksquare\neg p) \lor (\blacksquare\neg p \land \blacklozenge\neg p) \equiv_{\mathsf{S}_1^*} \blacksquare\neg p$,
  - $\psi = (\Box\blacksquare p \land \Diamond\blacksquare \bullet p) \lor (\Box \circ \blacklozenge p \land \Diamond \circ \blacklozenge \bullet p)$
    $\equiv_{\mathsf{S}_1^*} (\blacksquare p \land \blacksquare\neg p) \lor (\blacklozenge p \land \blacklozenge\neg p) \equiv_{\mathsf{S}_1^*} \blacklozenge p \land \blacklozenge\neg p$,
  - $\chi = (\Box\blacksquare p \land \Diamond\blacksquare \bullet p) \lor (\Box\blacksquare p \land \Box \circ \blacklozenge p)$
    $\equiv_{\mathsf{S}_1^*} (\blacksquare p \land \blacksquare\neg p) \lor (\blacksquare p \land \blacklozenge p) \equiv_{\mathsf{S}_1^*} \blacksquare p$,
  - $\neg\varphi = (\Box \circ \blacklozenge p \land \Diamond \circ \blacklozenge \bullet p) \lor (\Box\blacksquare p \land \Box \circ \blacklozenge p)$
    $\equiv_{\mathsf{S}_1^*} (\blacklozenge p \land \blacklozenge\neg p) \lor (\blacksquare p \land \blacklozenge p) \equiv_{\mathsf{S}_1^*} \blacklozenge p$,
  - $\neg\psi = (\Diamond\blacksquare \bullet p \land \Diamond \circ \blacklozenge \bullet p) \lor (\Box\blacksquare p \land \Box \circ \blacklozenge p)$
    $\equiv_{\mathsf{S}_1^*} (\blacksquare\neg p \land \blacklozenge\neg p) \lor (\blacksquare p \land \blacklozenge p) \equiv_{\mathsf{S}_1^*} \blacksquare p \lor \blacksquare\neg p$,
  - $\neg\chi = (\Diamond\blacksquare \bullet p \land \Diamond \circ \blacklozenge \bullet p) \lor (\Box \circ \blacklozenge p \land \Diamond \circ \blacklozenge \bullet p)$
    $\equiv_{\mathsf{S}_1^*} (\blacksquare\neg p \land \blacklozenge\neg p) \lor (\blacklozenge p \land \blacklozenge\neg p) \equiv_{\mathsf{S}_1^*} \blacklozenge\neg p$.

Logic-Sensitivity and Bitstring Semantics

Since $\blacksquare$ is a KD-modality in S_1^*, the Aristotelian diagram for $(\mathcal{H}_{\square\blacksquare}^{\circ\bullet}, \mathsf{S}_1^*)$ is a JSB hexagon, with pairwise contrarieties among $\blacksquare\neg p$, $\blacklozenge p \wedge \blacklozenge\neg p$ and $\blacksquare p$, i.e., among φ, ψ and χ, respectively; cf. Figure 5(b). The partition $\Pi_{\mathsf{S}_1^*}(\mathcal{H}_{\square\blacksquare}^{\circ\bullet})$ contains the following anchor formulas:

- $\lozenge\blacksquare\bullet p \wedge \lozenge\circ\blacklozenge\bullet p \equiv_{\mathsf{S}_1^*} \blacksquare\neg p \wedge \blacklozenge\neg p \equiv_{\mathsf{S}_1^*} \blacksquare\neg p$,
- $\square\circ\blacklozenge p \wedge \lozenge\circ\blacklozenge\bullet p \equiv_{\mathsf{S}_1^*} \blacklozenge p \wedge \blacklozenge\neg p$,
- $\square\blacksquare p \wedge \square\circ\blacklozenge p \equiv_{\mathsf{S}_1^*} \blacksquare p \wedge \blacklozenge p \equiv_{\mathsf{S}_1^*} \blacksquare p$.

This shows that the diagram for $(\mathcal{H}_{\square\blacksquare}^{\circ\bullet}, \mathsf{S}_1^*)$ is a length-3, i.e., *strong*, JSB hexagon. In comparison with $\Pi_{\mathsf{S}_0^*}(\mathcal{H}_{\square\blacksquare}^{\circ\bullet})$, note that the first anchor formula (viz., $\square\blacksquare p \wedge \lozenge\blacksquare\bullet p \equiv_{\mathsf{S}_1^*} \blacksquare p \wedge \blacksquare\neg p$) is S_1^*-inconsistent, and is thus absent from $\Pi_{\mathsf{S}_1^*}(\mathcal{H}_{\square\blacksquare}^{\circ\bullet})$.

- In S_2^*, the formulas of $\mathcal{H}_{\square\blacksquare}^{\circ\bullet}$ simplify as follows:

  - $\varphi = (\square\blacksquare p \wedge \lozenge\blacksquare\bullet p) \vee (\lozenge\blacksquare\bullet p \wedge \lozenge\circ\blacklozenge\bullet p)$
    $\equiv_{\mathsf{S}_2^*} (\square p \wedge \lozenge p) \vee (\lozenge p \wedge \lozenge\neg p) \equiv_{\mathsf{S}_2^*} \lozenge p$,
  - $\psi = (\square\blacksquare p \wedge \lozenge\blacksquare\bullet p) \vee (\square\circ\blacklozenge p \wedge \lozenge\circ\blacklozenge\bullet p)$
    $\equiv_{\mathsf{S}_2^*} (\square p \wedge \lozenge p) \vee (\square\neg p \wedge \lozenge\neg p) \equiv_{\mathsf{S}_2^*} \lozenge p \vee \lozenge\neg p$,
  - $\chi = (\square\blacksquare p \wedge \lozenge\blacksquare\bullet p) \vee (\square\blacksquare p \wedge \square\circ\blacklozenge p)$
    $\equiv_{\mathsf{S}_2^*} (\square p \wedge \lozenge p) \vee (\square p \wedge \square\neg p) \equiv_{\mathsf{S}_2^*} \square p$,
  - $\neg\varphi = (\square\circ\blacklozenge p \wedge \lozenge\circ\blacklozenge\bullet p) \vee (\square\blacksquare p \wedge \square\circ\blacklozenge p)$
    $\equiv_{\mathsf{S}_2^*} (\square\neg p \wedge \lozenge\neg p) \vee (\square p \wedge \square\neg p) \equiv_{\mathsf{S}_2^*} \square\neg p$,
  - $\neg\psi = (\lozenge\blacksquare\bullet p \wedge \lozenge\circ\blacklozenge\bullet p) \vee (\square\blacksquare p \wedge \square\circ\blacklozenge p)$
    $\equiv_{\mathsf{S}_2^*} (\lozenge p \wedge \lozenge\neg p) \vee (\square p \wedge \square\neg p) \equiv_{\mathsf{S}_2^*} \square p \wedge \square\neg p$,
  - $\neg\chi = (\lozenge\blacksquare\bullet p \wedge \lozenge\circ\blacklozenge\bullet p) \vee (\square\circ\blacklozenge p \wedge \lozenge\circ\blacklozenge\bullet p)$
    $\equiv_{\mathsf{S}_2^*} (\lozenge p \wedge \lozenge\neg p) \vee (\square\neg p \wedge \lozenge\neg p) \equiv_{\mathsf{S}_2^*} \lozenge\neg p$.

Since $\square$ is a KF-modality in S_2^*, the Aristotelian diagram for $(\mathcal{H}_{\square\blacksquare}^{\circ\bullet}, \mathsf{S}_2^*)$ is a JSB hexagon, with pairwise contrarieties among $\lozenge p$, $\square p \wedge \square\neg p$ and $\lozenge\neg p$, i.e., among φ, $\neg\psi$ and $\neg\chi$, respectively; cf. Figure 5(c). The partition $\Pi_{\mathsf{S}_2^*}(\mathcal{H}_{\square\blacksquare}^{\circ\bullet})$ contains the following anchor formulas:

- $\square\blacksquare p \wedge \lozenge\blacksquare\bullet p \equiv_{\mathsf{S}_2^*} \square p \wedge \lozenge p \equiv_{\mathsf{S}_2^*} \lozenge p$
- $\square\circ\blacklozenge p \wedge \lozenge\circ\blacklozenge\bullet p \equiv_{\mathsf{S}_2^*} \square\neg p \wedge \lozenge\neg p \equiv_{\mathsf{S}_2^*} \lozenge\neg p$
- $\square\blacksquare p \wedge \square\circ\blacklozenge p \equiv_{\mathsf{S}_2^*} \square p \wedge \square\neg p$

This shows that the diagram for $(\mathcal{H}_{\square\blacksquare}^{\circ\bullet}, \mathsf{S}_2^*)$ is a length-3, i.e., *strong*, JSB hexagon. In comparison with $\Pi_{\mathsf{S}_0^*}(\mathcal{H}_{\square\blacksquare}^{\circ\bullet})$, note that the second anchor formula (viz., $\Diamond\blacksquare\bullet p \wedge \Diamond\circ\blacklozenge\bullet p \equiv_{\mathsf{S}_2^*} \Diamond p \wedge \Diamond\neg p$) is S_2^*-inconsistent, and is thus absent from $\Pi_{\mathsf{S}_2^*}(\mathcal{H}_{\square\blacksquare}^{\circ\bullet})$.

- In S_3^*, the formulas of $\mathcal{H}_{\square\blacksquare}^{\circ\bullet}$ simplify as follows:

  - $\varphi = (\square\blacksquare p \wedge \Diamond\blacksquare\bullet p) \vee (\Diamond\blacksquare\bullet p \wedge \Diamond\circ\blacklozenge\bullet p)$
    $\equiv_{\mathsf{S}_3^*} (\blacksquare p \wedge \blacksquare\neg p) \vee (\blacksquare\neg p \wedge \blacklozenge\neg p) \equiv_{\mathsf{S}_3^*} \blacksquare\neg p$,
  - $\psi = (\square\blacksquare p \wedge \Diamond\blacksquare\bullet p) \vee (\square\circ\blacklozenge p \wedge \Diamond\circ\blacklozenge\bullet p)$
    $\equiv_{\mathsf{S}_3^*} (\blacksquare p \wedge \blacksquare\neg p) \vee (\blacklozenge p \wedge \blacklozenge\neg p) \equiv_{\mathsf{S}_3^*} \blacksquare p \wedge \blacksquare\neg p$,
  - $\chi = (\square\blacksquare p \wedge \Diamond\blacksquare\bullet p) \vee (\square\blacksquare p \wedge \square\circ\blacklozenge p)$
    $\equiv_{\mathsf{S}_3^*} (\blacksquare p \wedge \blacksquare\neg p) \vee (\blacksquare p \wedge \blacklozenge p) \equiv_{\mathsf{S}_3^*} \blacksquare p$,
  - $\neg\varphi = (\square\circ\blacklozenge p \wedge \Diamond\circ\blacklozenge\bullet p) \vee (\square\blacksquare p \wedge \square\circ\blacklozenge p)$
    $\equiv_{\mathsf{S}_3^*} (\blacklozenge p \wedge \blacklozenge\neg p) \vee (\blacksquare p \wedge \blacklozenge p) \equiv_{\mathsf{S}_3^*} \blacklozenge p$,
  - $\neg\psi = (\Diamond\blacksquare\bullet p \wedge \Diamond\circ\blacklozenge\bullet p) \vee (\square\blacksquare p \wedge \square\circ\blacklozenge p)$
    $\equiv_{\mathsf{S}_3^*} (\blacksquare\neg p \wedge \blacklozenge\neg p) \vee (\blacksquare p \wedge \blacklozenge p) \equiv_{\mathsf{S}_3^*} \blacklozenge p \vee \blacklozenge\neg p$,
  - $\neg\chi = (\Diamond\blacksquare\bullet p \wedge \Diamond\circ\blacklozenge\bullet p) \vee (\square\circ\blacklozenge p \wedge \Diamond\circ\blacklozenge\bullet p)$
    $\equiv_{\mathsf{S}_3^*} (\blacksquare\neg p \wedge \blacklozenge\neg p) \vee (\blacklozenge p \wedge \blacklozenge\neg p) \equiv_{\mathsf{S}_3^*} \blacklozenge\neg p$.

Since $\blacksquare$ is a KF-modality in S_3^*, the Aristotelian diagram for $(\mathcal{H}_{\square\blacksquare}^{\circ\bullet}, \mathsf{S}_3^*)$ is a JSB hexagon, with pairwise contrarieties among $\blacklozenge p$, $\blacksquare p \wedge \blacksquare\neg p$ and $\blacklozenge\neg p$, i.e., among $\neg\varphi$, ψ and $\neg\chi$, respectively; cf. Figure 5(d). The partition $\Pi_{\mathsf{S}_3^*}(\mathcal{H}_{\square\blacksquare}^{\circ\bullet})$ contains the following anchor formulas:

  - $\square\blacksquare p \wedge \Diamond\blacksquare\bullet p \equiv_{\mathsf{S}_3^*} \blacksquare p \wedge \blacksquare\neg p$,
  - $\Diamond\blacksquare\bullet p \wedge \Diamond\circ\blacklozenge\bullet p \equiv_{\mathsf{S}_3^*} \blacksquare\neg p \wedge \blacklozenge\neg p \equiv_{\mathsf{S}_3^*} \blacklozenge\neg p$,
  - $\square\blacksquare p \wedge \square\circ\blacklozenge p \equiv_{\mathsf{S}_3^*} \blacksquare p \wedge \blacklozenge p \equiv_{\mathsf{S}_3^*} \blacklozenge p$.

This shows that the diagram for $(\mathcal{H}_{\square\blacksquare}^{\circ\bullet}, \mathsf{S}_3^*)$ is a length-3, i.e., *strong*, JSB hexagon. In comparison with $\Pi_{\mathsf{S}_0^*}(\mathcal{H}_{\square\blacksquare}^{\circ\bullet})$, note that the third anchor formula (viz., $\square\circ\blacklozenge p \wedge \Diamond\circ\blacklozenge\bullet p \equiv_{\mathsf{S}_3^*} \blacklozenge p \wedge \blacklozenge\neg p$) is S_3^*-inconsistent, and is thus absent from $\Pi_{\mathsf{S}_3^*}(\mathcal{H}_{\square\blacksquare}^{\circ\bullet})$.

- In S_4^*, the formulas of $\mathcal{H}_{\square\blacksquare}^{\circ\bullet}$ simplify as follows:

  - $\varphi = (\square\blacksquare p \wedge \Diamond\blacksquare\bullet p) \vee (\Diamond\blacksquare\bullet p \wedge \Diamond\circ\blacklozenge\bullet p)$
    $\equiv_{\mathsf{S}_4^*} (\square p \wedge \Diamond p) \vee (\Diamond p \wedge \Diamond\neg p) \equiv_{\mathsf{S}_4^*} \Diamond p$,

- $\psi = (\Box\blacksquare p \wedge \Diamond\blacksquare\bullet p) \vee (\Box\circ\blacklozenge p \wedge \Diamond\circ\blacklozenge\bullet p)$
  $\equiv_{\mathsf{S}_4^*} (\Box p \wedge \Diamond p) \vee (\Box\neg p \wedge \Diamond\neg p) \equiv_{\mathsf{S}_4^*} \Box p \vee \Box\neg p,$
- $\chi = (\Box\blacksquare p \wedge \Diamond\blacksquare\bullet p) \vee (\Box\blacksquare p \wedge \Box\circ\blacklozenge p)$
  $\equiv_{\mathsf{S}_4^*} (\Box p \wedge \Diamond p) \vee (\Box p \wedge \Box\neg p) \equiv_{\mathsf{S}_4^*} \Box p,$
- $\neg\varphi = (\Box\circ\blacklozenge p \wedge \Diamond\circ\blacklozenge\bullet p) \vee (\Box\blacksquare p \wedge \Box\circ\blacklozenge p)$
  $\equiv_{\mathsf{S}_4^*} (\Box\neg p \wedge \Diamond\neg p) \vee (\Box p \wedge \Box\neg p) \equiv_{\mathsf{S}_4^*} \Box\neg p,$
- $\neg\psi = (\Diamond\blacksquare\bullet p \wedge \Diamond\circ\blacklozenge\bullet p) \vee (\Box\blacksquare p \wedge \Box\circ\blacklozenge p)$
  $\equiv_{\mathsf{S}_4^*} (\Diamond p \wedge \Diamond\neg p) \vee (\Box p \wedge \Box\neg p) \equiv_{\mathsf{S}_4^*} \Diamond p \wedge \Diamond\neg p,$
- $\neg\chi = (\Diamond\blacksquare\bullet p \wedge \Diamond\circ\blacklozenge\bullet p) \vee (\Box\circ\blacklozenge p \wedge \Diamond\circ\blacklozenge\bullet p)$
  $\equiv_{\mathsf{S}_4^*} (\Diamond p \wedge \Diamond\neg p) \vee (\Box\neg p \wedge \Diamond\neg p) \equiv_{\mathsf{S}_4^*} \Diamond\neg p.$

Since $\Box$ is a KD-modality in S_4^*, the Aristotelian diagram for $(\mathcal{H}_{\Box\blacksquare}^{\circ\bullet}, \mathsf{S}_4^*)$ is a JSB hexagon, with pairwise contrarieties among $\Box\neg p$, $\Diamond p \wedge \Diamond\neg p$ and $\Box p$, i.e., among $\neg\varphi$, $\neg\psi$ and χ, respectively; cf. Figure 5(e). The partition $\Pi_{\mathsf{S}_4^*}(\mathcal{H}_{\Box\blacksquare}^{\circ\bullet})$ contains the following anchor formulas:

- $\Box\blacksquare p \wedge \Diamond\blacksquare\bullet p \equiv_{\mathsf{S}_4^*} \Box p \wedge \Diamond p \equiv_{\mathsf{S}_4^*} \Box p$
- $\Diamond\blacksquare\bullet p \wedge \Diamond\circ\blacklozenge\bullet p \equiv_{\mathsf{S}_4^*} \Diamond p \wedge \Diamond\neg p$
- $\Box\circ\blacklozenge p \wedge \Diamond\circ\blacklozenge\bullet p \equiv_{\mathsf{S}_4^*} \Box\neg p \wedge \Diamond\neg p \equiv_{\mathsf{S}_4^*} \Box\neg p$

This shows that the diagram for $(\mathcal{H}_{\Box\blacksquare}^{\circ\bullet}, \mathsf{S}_4^*)$ is a length-3, i.e., *strong*, JSB hexagon. In comparison with $\Pi_{\mathsf{S}_0^*}(\mathcal{H}_{\Box\blacksquare}^{\circ\bullet})$, note that the fourth anchor formula (viz., $\Box\blacksquare p \wedge \Box\circ\blacklozenge p \equiv_{\mathsf{S}_4^*} \Box p \wedge \Box\neg p$) is S_4^*-inconsistent, and is thus absent from $\Pi_{\mathsf{S}_4^*}(\mathcal{H}_{\Box\blacksquare}^{\circ\bullet})$.

These considerations jointly show that $\langle \mathcal{L}_{\Box\blacksquare}^{\circ\bullet}, \mathcal{H}_{\Box\blacksquare}^{\circ\bullet}, \mathsf{S}_0^*, \mathsf{S}_1^*, \mathsf{S}_2^*, \mathsf{S}_3^*, \mathsf{S}_4^* \rangle$ constitutes a *complete* solution to Open Problem 2. We have thus succeeded in 're-telling the story' from Section 3, but now from a semantic instead of a purely combinatorial perspective (i.e., in terms of logical systems instead of bitstrings).

6 Conclusion

In this paper we have continued our exploration of the interface between logic-sensitivity and bitstring semantics in Aristotelian diagrams. This research line was initiated in [17], where it remained limited to squares of opposition. However, we have argued in this paper that there are good reasons for broadening the scope from squares to hexagons of opposition. This generalization comes

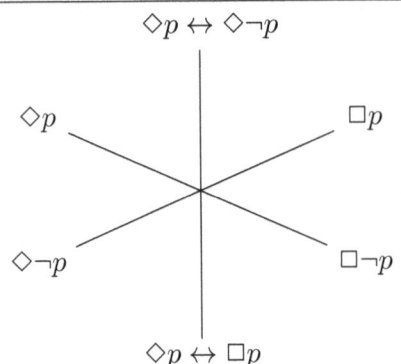

(a) Length-4 U12 hexagon for $(\mathcal{H}^{\circ\bullet}_{\square\blacksquare}, \mathsf{S}^*_0)$, after simplifying the formulas in $\mathcal{H}^{\circ\bullet}_{\square\blacksquare}$.

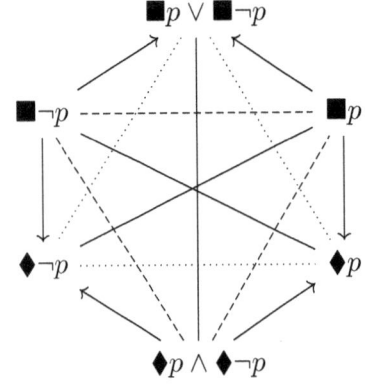

(b) Strong JSB hexagon for $(\mathcal{H}^{\circ\bullet}_{\square\blacksquare}, \mathsf{S}^*_1)$, after simplifying.

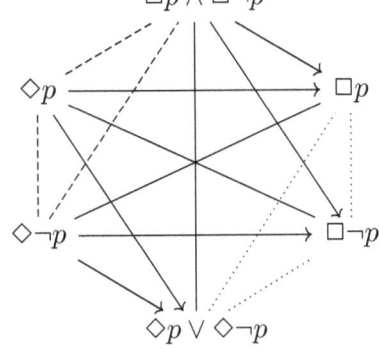

(c) Strong JSB hexagon for $(\mathcal{H}^{\circ\bullet}_{\square\blacksquare}, \mathsf{S}^*_2)$, after simplifying.

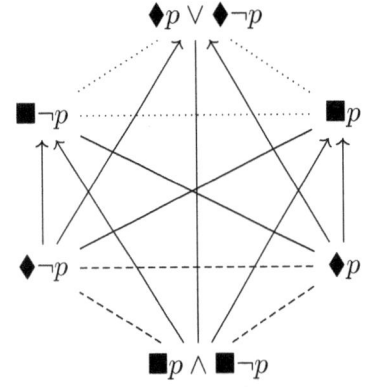

(d) Strong JSB hexagon for $(\mathcal{H}^{\circ\bullet}_{\square\blacksquare}, \mathsf{S}^*_3)$, after simplifying.

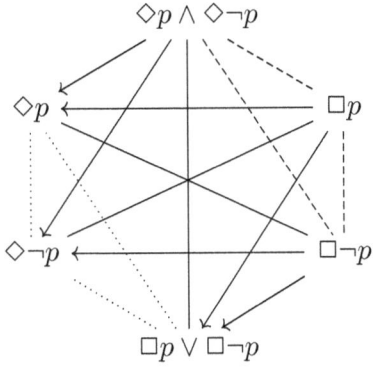

(e) Strong JSB hexagon for $(\mathcal{H}^{\circ\bullet}_{\square\blacksquare}, \mathsf{S}^*_4)$, after simplifying.

Figure 5: Five Aristotelian hexagons for the solution to Open Problem 2.

about very naturally, although it does entail that we have to take into account a new layer of complexity, viz., the existence of Boolean subtypes in hexagons (and larger diagrams) of opposition. Furthermore, we have argued that the interaction between logic-sensitivity and bitstring semantics continues to be very elegant at the purely combinatorial level, yet less satisfactory at the semantic level. In particular, we have shown how the (complete but *ad hoc*) solution to Open Problem 1 from [17] can be generalized to a (still complete, but also still *ad hoc*) solution to Open Problem 2.

Acknowledgments

This research was funded by the ID-N project *BITSHARE: Bitstring Semantics for Human and Artificial Reasoning* (IDN-19-009, Internal Funds, KU Leuven). The first author holds a research professorship (BOFZAP) at KU Leuven. Both authors would like to thank Hans Smessaert for his feedback on an earlier version of this paper.

References

[1] BÉZIAU, J.-Y. New light on the square of oppositions and its nameless corner. *Logical Investigations 10* (2003), 218–232.

[2] BÉZIAU, J.-Y., AND BASTI, G., Eds. *The Square of Opposition: A Cornerstone of Thought*. Springer, 2017.

[3] BÉZIAU, J.-Y., AND JACQUETTE, D., Eds. *Around and Beyond the Square of Opposition*. Springer, 2012.

[4] BEZIAU, J.-Y., AND VANDOULAKIS, I. M., Eds. *The Exoteric Square of Opposition*. Springer, 2022.

[5] BLANCHÉ, R. Sur l'opposition des concepts. *Theoria 19* (1953), 89–130.

[6] BLANCHÉ, R. *Structures Intellectuelles*. Vrin, 1966.

[7] CIUCCI, D., DUBOIS, D., AND PRADE, H. Structures of opposition induced by relations. the Boolean and the gradual cases. *Annals of Mathematics and Artificial Intelligence 76* (2016), 351–373.

[8] DE KLERCK, A., AND DEMEY, L. Alpha-structures and ladders in logical geometry. *Studia Logica* (forthcoming, doi:10.1007/s11225-024-10142-0).

[9] DE KLERCK, A., VIGNERO, L., AND DEMEY, L. Morphisms between Aristotelian diagrams. *Logica Universalis 18* (2023), 49–83.

[10] DE KLERCK, A., VIGNERO, L., AND DEMEY, L. Category theory for Aristotelian diagrams: The debate on singular propositions. In *Diagrammatic Representation and Inference*, J. Lemanski et al., Eds., LNCS 14981. Springer, 2024, pp. 153–161.

[11] DEMEY, L. Interactively illustrating the context-sensitivity of Aristotelian diagrams. In *Modeling and Using Context*, H. Christiansen, I. Stojanovic, and G. Papadopoulos, Eds., LNCS 9405. Springer, 2015, pp. 331–345.

[12] DEMEY, L. Computing the maximal Boolean complexity of families of Aristotelian diagrams. *Journal of Logic and Computation 28* (2018), 1323–1339.

[13] DEMEY, L. Boolean considerations on John Buridan's octagons of oppositions. *History and Philosophy of Logic 40* (2019), 116–134.

[14] DEMEY, L. Metalogic, metalanguage and logical geometry. *Logique et Analyse 248* (2019), 453–478.

[15] DEMEY, L. From Euler diagrams in Schopenhauer to Aristotelian diagrams in logical geometry. In *Language, Logic, and Mathematics in Schopenhauer*, J. Lemanski, Ed. Springer, 2020, pp. 181–205.

[16] DEMEY, L. Logic-sensitivity of Aristotelian diagrams in non-normal modal logics. *Axioms 10*, 128 (2021), 1 – 25.

[17] DEMEY, L., AND FRIJTERS, S. Logic-sensitivity and bitstring semantics in the square of opposition. *Journal of Philosophical Logic 52* (2023), 1703–1721.

[18] DEMEY, L., AND SMESSAERT, H. Logical and geometrical distance in polyhedral Aristotelian diagrams in knowledge representation. *Symmetry 9(10)*, 204 (2017).

[19] DEMEY, L., AND SMESSAERT, H. Combinatorial bitstring semantics for arbitrary logical fragments. *Journal of Philosophical Logic 47* (2018), 325–363.

[20] FRIJTERS, S. Generalizing Aristotelian relations and diagrams. In *Diagrammatic Representation and Inference*, V. Giardino, S. Linker, R. Burns, F. Bellucci, J.-M. Boucheix, and P. Viana, Eds., LNCS 13462. Springer, 2022, pp. 329 – 337.

[21] FRIJTERS, S., AND DEMEY, L. The modal logic of Aristotelian diagrams. *Axioms 12*, 471 (2023), 1–26.

[22] GEUDENS, C., AND DEMEY, L. On the Aristotelian roots of the modal square of opposition. *Logique et Analyse 255* (2021), 313 – 348.

[23] GEUDENS, C., AND DEMEY, L. *The Modal Logic of John Fabri of Valenciennes (c. 1500). A Study in Token-Based Semantics*. Springer, 2022.

[24] JACOBY, P. A triangle of opposites for types of propositions in Aristotelian logic. *New Scholasticism 24* (1950), 32–56.

[25] JACOBY, P. Contrariety and the triangle of opposites in valid inferences. *New Scholasticism 34* (1960), 141–169.

[26] LEMANSKI, J., AND DEMEY, L. Schopenhauer's partition diagrams and logical geometry. In *Diagrammatic Representation and Inference*, A. Basu, G. Stapleton, S. Linker, C. Legg, E. Manalo, and P. Viana, Eds., LNCS 12909. Springer, 2021, pp. 149–165.

[27] PELLISSIER, R. Setting n-opposition. *Logica Universalis 2* (2008), 235–263.

[28] PIZZI, C. Generalization and composition of modal squares of opposition. *Logica Universalis 10* (2016), 313–325.

[29] READ, S. Aristotle and Łukasiewicz on existential import. *Journal of the Amer-*

ican Philosophical Association **1** (2015), 535–544.

[30] SESMAT, A. *Logique II. Les Raisonnements. La syllogistique.* Hermann, 1951.

[31] SMESSAERT, H. On the 3D visualisation of logical relations. *Logica Universalis 3* (2009), 303–332.

[32] SMESSAERT, H., AND DEMEY, L. Logical and geometrical complementarities between Aristotelian diagrams. In *Diagrammatic Representation and Inference*, T. Dwyer, H. Purchase, and A. Delaney, Eds., LNCS 8578. Springer, 2014, pp. 246–260.

[33] SMESSAERT, H., AND DEMEY, L. Logical geometries and information in the square of opposition. *Journal of Logic, Language and Information 23* (2014), 527–565.

[34] SMESSAERT, H., AND DEMEY, L. The unreasonable effectiveness of bitstrings in logical geometry. In *The Square of Opposition: A Cornerstone of Thought*, J.-Y. Béziau and G. Basti, Eds. Springer, 2017, pp. 197–214.

[35] SMESSAERT, H., AND DEMEY, L. On the logical geometry of geometric angles. *Logica Universalis 16* (2022), 581–601.

[36] SMESSAERT, H., AND DEMEY, L. Aristotelian diagrams for the proportional quantifier 'most'. *Axioms 12*, 3 (2023), 1–13.

[37] VIGNERO, L. Combining and relating Aristotelian diagrams. In *Diagrammatic Representation and Inference* (2021), A. Basu, G. Stapleton, S. Linker, C. Legg, E. Manalo, and P. Viana, Eds., Springer, pp. 221–228.